Heat Pumps

RENEWABLE ENERGIES SERIES

skills2learn
www.skills2learn.com
Experts in e-learning & virtual reality simulation

CENGAGE
Learning·

Australia • Brazil • Japan • Korea • Mexico • Singapore • Spain • United Kingdom • United States

Heat Pumps, 1st Edition
Skills2Learn

Publishing Director: Linden Harris

Commissioning Editor: Lucy Mills

Development Editor: Lauren Darby

Senior Project Editor: Alison Burt

Senior Manufacturing Buyer: Eyvett Davis

Typesetter: MPS Limited

Cover design: HCT Creative

For product information and technology assistance,
contact **emea.info@cengage.com.**

For permission to use material from this text or product,
and for permission queries,
email **emea.permissions@cengage.com.**

British Library Cataloguing-in-Publication Data

A catalogue record for this book is available from the British Library.

ISBN: 978-1-4080-6466-5

Cengage Learning EMEA

Cheriton House, North Way, Andover, Hampshire, SP10 5BE
United Kingdom

Cengage Learning products are represented in Canada by Nelson Education Ltd.

For your lifelong learning solutions, visit **www.cengage.co.uk**

Purchase your next print book, e-book or e-chapter at **www.cengagebrain.com**

Printed in China by RR Donnelley

1 2 3 4 5 6 7 8 9 10 – 15 14 13

Contents

Foreword

The energy sector is a significant part of the UK economy and a major employer of people. It has a huge impact on the environment and plays a massive role in our everyday life, shaping both our work and domestic habits and processes. With environmental issues such as climate change and sustainable sourcing of energy now playing an important role in our society, there is a need to educate a significant pool of people about the future technologies with renewable energies in all likelihood playing an increasingly significant part in our total energy requirements.

This environmental and renewable energy series of e-learning programmes and text workbooks has been developed to provide a structured blended learning approach that will enhance the learning experience and stimulate a deeper understanding of the renewable energy trades and give an awareness of sustainability issues. The content within these learning materials has been aligned as far as is currently possible to the units of the National Occupational Standards and can be used as a support tool whilst studying for any relevant vocational qualifications.

The uniqueness of this renewable energy series is that it aims to bridge the gap between classroom-based and practical-based learning. The workbooks provide classroom-based activities that can involve learners in discussions and research tasks as well as providing them with understanding and knowledge of the subject. The e-learning programmes take the subject further, with high quality images, animations and audio further enhancing the content and showing information in a different light. In addition, the e-practical side of the e-learning places the learner in a virtual environment where they can move around freely, interact with objects and use the knowledge and skills they have gained from the workbook and e-learning to complete a set of tasks whilst in the comfort of a safe working environment.

The workbooks and e-learning programmes are designed to help learners continuously improve their skills and provide a confident and sound knowledge base before getting their hands dirty in the real world.

About e-Consortia

This series of renewable energy workbooks and e-learning pro-grammes has been developed by the E-Renewable Consortium. The consortium is a group of colleges and organizations that are passion-ate about the renewable energy industry and are determined to enhance the learning experiences of people within the different trades or those that are new to it.

The consortium members have many years experience in the renew-able energy and educational sectors and have created this blended learning approach of interactive e-learning programmes and text workbooks to achieve the aim of:

- Providing accessible training in different areas of renewable energy
- Bridging the gap between classroom-based and practical-based learning
- Providing a concentrated set of improvement learning modules
- Enabling learners to gain new skills and qualifications more effectively
- Improving functional skills and awareness of sustainability issues within the industry
- Promoting health and safety in the industry
- Encouraging training and continuous professional development.

For more information about this renewable energy series please visit: **http://skills2learn.cengage.co.uk/9-renewable-energy**

About e-learning

INTRODUCTION

This renewable energies series of workbooks and e-learning programmes uses a blended learning approach to train learners about renewable energy skills. Blended learning allows training to be delivered through different mediums such as books, e-learning (computer-based training), practical workshops, and traditional classroom techniques. These training methods are designed to complement each other and work in tandem to achieve overall learning objectives and outcomes.

E-LEARNING

The Heat Pumps e-learning programme that is also available to sit alongside this workbook offers a different method of learning. With technology playing an increasingly important part of everyday life, e-learning uses visually rich 2D and 3D graphics/animation, audio, video, text and interactive quizzes, to allow you to engage with the content and learn at your own pace and in your own time.

E-ASSESSMENT

Part of the e-learning programme is an e-assessment "End test". This facility allows you to be self-tested using interactive multimedia by answering questions on the e-learning modules you will have covered in the programme. The e-assessment provides feedback on both correctly and incorrectly answered questions. If answers are incorrect the learner is advised to revisit the learning materials they need to study further.

E-PRACTICAL

Part of the e-learning programme is an e-practical interactive scenario. This facility allows you to be immersed in a virtual reality situation where the choices you make affect the outcome. Using 3D technology, you can move freely around the environment, interact with objects, carry out tests, and make decisions and mistakes until you have mastered the subject. By practising in a virtual environment you will not only be able to see what you've learnt but also analyze your approach and thought process to the problem.

BENEFITS OF E-LEARNING

Diversity – E-learning can be used for almost anything. With the correct approach any subject can be brought to life to provide an interactive training experience.

Technology – Advancements in computer technology now allow a wide range of spectacular and engaging e-learning to be delivered to a wider population.

Captivate and Motivate – Hold the learner's attention for longer with the use of high quality graphics, animation, sound and interactivity.

Safe Environment – E-Practical scenarios can create environments which simulate potentially harmful real-life situations or replicate a piece of dangerous equipment, therefore allowing the learner to train and gain experience and knowledge in a completely safe environment.

Instant Feedback – Learners can undertake training assessments which feedback results instantly. This can provide information on where they need to re-study or congratulate them on passing the assessment. Results and Certificates could also be printed for future records.

On-Demand – Can be accessed 24 hours a day, 7 days a week, 365 days of the year. You can access the content at any time and view it at your own pace.

Portable Solutions – Can be delivered via a CD, website or LMS. Learners no longer need to travel to all lectures, conferences, meetings or training days. This saves many man-hours in reduced travelling, cost of hotels and expenses amongst other things.

Reduction of Costs – Can be used to teach best practice processes on jobs which use large quantities of expensive materials. Learners can practice their techniques and boost their confidence to a high enough standard before being allowed near real materials.

HEAT PUMPS E-LEARNING

The aim of the heat pumps e-learning programme is to enhance a learner's knowledge and understanding of heat pumps installation and systems. The course content is aligned to units from the Environmental National Occupational Standards (NOS) so can be used for study towards certification.

The programme gives the learners an understanding of the different types of heat pumps, as well as looking at sustainability, health and safety and functional skills in an interactive and visually engaging manner. It also provides a 'real-life' scenario where the learner can apply the knowledge gained from the tutorials in a safe yet practical way.

By using and completing this programme, it is expected that learners will:

- Explain what a heat pump is and how to calculate the efficiency of a heat pump
- List the unit components of a heat pump
- Explain how a ground source heat collector works
- Explain how an air source heat collector works
- List the principles of heat pump selection and system design
- Explain the preparatory work required for installation and know how to install a heat pump
- List the requirements and know how to test, commission and handover heat pump installations

The e-learning programme is divided into the following learning modules:

- Getting Started
- Heat Pumps Overview and Considerations
- Health and Safety
- Heat Pump Types

- Ground Source Collectors
- Air Source Collectors
- System Design
- Installation
- Testing, Commissioning and Handover
- Interactive E-Practical Scenarios

THE RENEWABLE ENERGIES SERIES

As part of the renewable energies series the following e-learning programmes and workbooks are available. For more information please visit: **http://skills2learn.cengage.co.uk/9-renewable-energy**

- Introduction to Renewable Energies
- Solar Thermal Hot Water
- Solar PV
- Building Heat Loss Calculator (programme only)
- Solar Radiation Calculator (programme only)

About the NOS

The National Occupational Standards (NOS) provide a framework of information that outline the skills, knowledge and understanding required to carry out work-based activities within a given vocation. Each standard is divided into units that cover specific activities of that occupation. Employers, employees, teachers and learners can use these standards as an information, support and reference resource that will enable them to understand the skills and criteria required for good practice in the workplace.

The standards are used as a basis to develop many vocational qualifications in the United Kingdom for a wide range of occupations. This workbook and associated e-learning programme are aligned to the Environmental National Occupational Standards (NOS) by being designed against the Qualification and Credit Framework (QCF) units, which are developed from the NOS. Such a process is a requirement of the minimum technical competency (MTC) document for heat pump installers. Therefore this book and its associated e-learning support full training towards a certificate as recognized by all bodies offering routes to the Microgeneration Certification Scheme (MCS) as evidence of suitable training. The information within relates to the following units:

- Know the health and safety risks and safe systems of work associated with heat pump installation work
- Know the requirements of relevant regulations/standards relating to practical installation, testing and commissioning activities for heat pump installation work
- Know the components and operational characteristics of heat pump units
- Know the different types of heat pump units and system arrangements for hydraulic emitter circuits
- Know the fundamental principles of heat pump selection and system design that are common to both air and ground source heat pumps

- Know the fundamental design principles for ground source 'closed loop' heat pump collector circuit design and component sizing
- Know the layouts of 'open loop' water filled heat pump collector circuits
- Know the fundamental design considerations that are specific to air source heat pumps
- Know the preparatory work required and be able to undertake the preparatory work required for heat pump installation work
- Know the requirements to install heat pump units
- Be able to install air and ground source heat pump units
- Know how to and be able to test heat pump system installations
- Know the requirements to commission heat pump system installations and be able to commission heat pump system installations
- Know the requirements to handover heat pump system installations and be able to handover heat pump system installations

SUMMARY OF THE ABOVE

Or in simplified English, this book and the e-learning training materials have been designed around the latest guidance from all the relevant bodies that will support a full heat pump training course and assessment process.

They share the knowledge in a simple-to-digest format that is more enjoyable to use and so more likely to be successful in sharing the important information required to install, commission, handover and maintain heat pump systems.

About the book

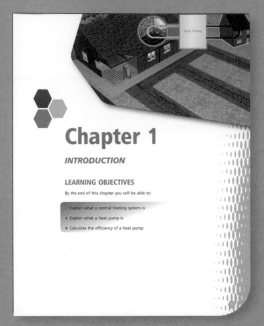

Learning Objectives at the start of each chapter explain the skills and knowledge you need to be proficient in and understand by the end of the chapter.

Activities are practical tasks that engage you in the subject and further your understanding.

E-learning Icons link the workbook content to the e-learning programme.

Health and Safety Boxes draw attention to essential health and safety information.

Note on UK Standards draws your attention to relevant building regulations.

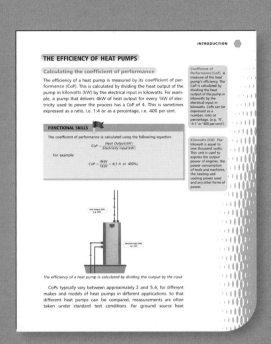

Functional Skills Icons highlight activities that develop and test your Maths, English and ICT key skills.

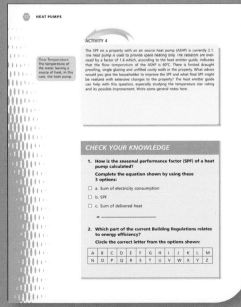

Check Your Knowledge at the end of each chapter to test your knowledge and understanding.

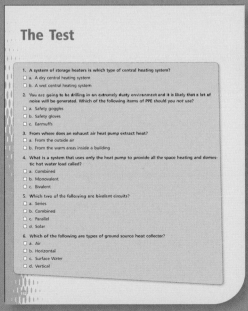

End Test in Chapter 9 checks your knowledge on all the information within the workbook.

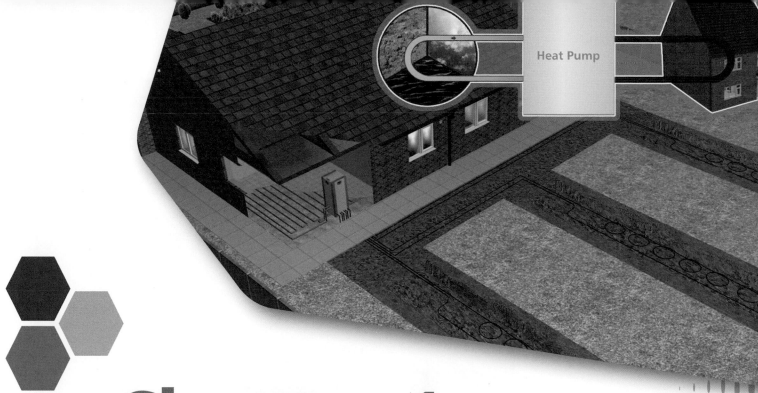

Heat Pump

Chapter 1

INTRODUCTION

LEARNING OBJECTIVES

By the end of this chapter you will be able to:

- Explain what a central heating system is

- Explain what a heat pump is

- Calculate the efficiency of a heat pump

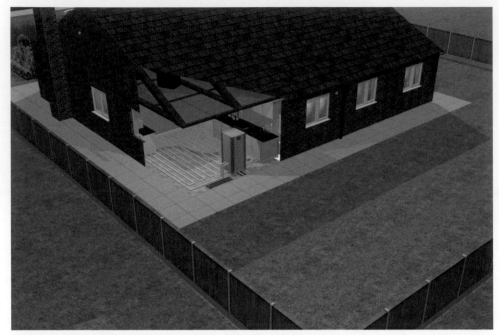

Cross section of house with heat pump installed

CENTRAL HEATING SYSTEMS

An introduction to central heating

Domestic Hot Water (DHW) Water which is heated and supplied for washing and bathing via taps or showers. DHW is always potable water and should be handled carefully to manage both Legionella and scalding risks.

A central heating system is generally defined as being a system that supplies heat from a single source to every room in a house instead of relying on an individual heat source in every room. Central heating is more effective as well as being relatively easy to maintain, control, clean and service. It can also provide both **domestic hot water** and space heating.

Central heating systems are divided into two basic types; dry and wet systems. In a dry system, heated air provides the warmth for the rooms in the house. In a wet system, heated water provides the heat.

It is now increasingly common for two or more heat sources to be used within the central heating system, for example, a mixed gas boiler and heat pump heating system providing the heat for the property. A mixed heating system such as this is called a bivalent (two heat sources) or multivalent (more than two heat sources) heating system.

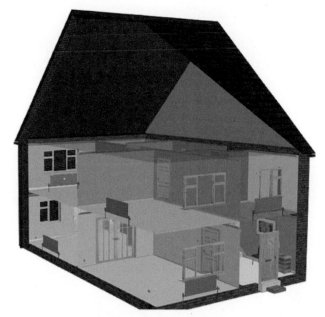

House with central heating system

Dry central heating systems

There are three main types of dry central heating system:

- Ducted warm air systems
- Electrical underfloor and ceiling systems
- Storage heaters

Ducted warm air systems

In a ducted warm air system air is warmed by being passed, relatively quickly, over a metal **heat exchanger**. A fan then blows the warm air through ducts into the rooms to be heated. A ducted warm air system is the only dry heating system that can truly be described as central because there is one central heating source, usually a gas-fired boiler.

> **Heat Exchanger** A heat exchanger is a device designed to efficiently transfer heat from one medium to another.

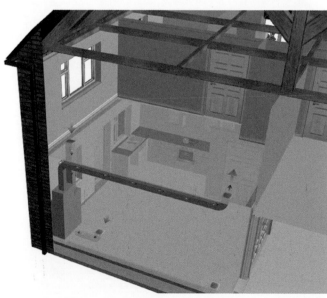

A ducted warm air system

Electrical underfloor and ceiling systems

Electrical underfloor and ceiling heating systems use elements built into the structure of the floor or ceiling to warm the surface and radiate heat into the room.

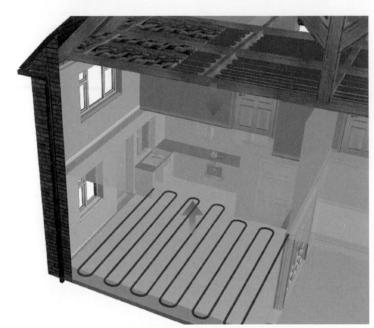

An electrical underfloor and ceiling system

Storage heaters

A system of individual storage heaters is also classed as a central heating system. The heaters contain heat-retaining firebricks which are heated during the night by low cost electricity. This heat is then released into the rooms during the day.

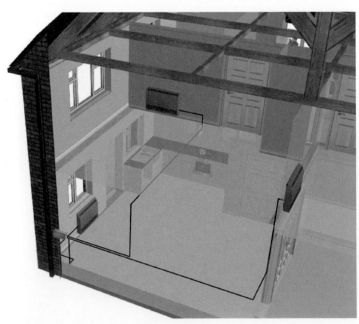

Storage heaters

E-LEARNING

Use the e-learning programme to see how the different dry central heating systems work.

Wet central heating systems

In wet central heating systems water is heated by a boiler which can be run on natural gas, bottled gas, oil, electricity or solid fuel. The water is then pumped through small bore pipes to radiators, convector heaters or into wet underfloor heating where the heat from the water is released into the room. The water is then circulated back to the boiler for reheating. This type of system can also be used to heat the domestic hot water supply as well as the rooms in the house.

E-LEARNING

Use the e-learning programme to see a demonstration on how wet central heating systems work.

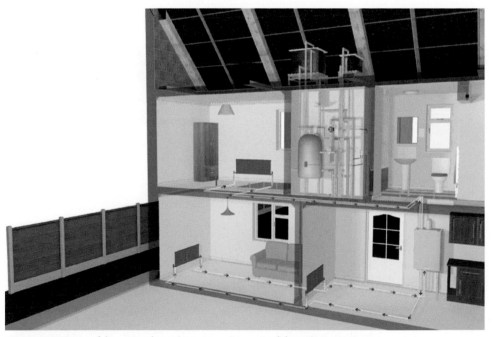

Cross section of house showing a wet central heating system

HEAT PUMPS

What is a heat pump?

A heat pump is a device for transferring heat from a lower temperature **heat source** to a higher temperature heat sink. Heat pumps use a refrigeration cycle to extract low temperature heat from a source, upgrade the heat to a higher, more useful temperature and then distribute this higher temperature heat into a heat sink.

A heat pump can therefore replace the boiler in a central heating system. A well-designed and properly installed heat pump is cheaper to run, produces less carbon and is more efficient than a boiler. The source of the low temperature heat can be air, water or the ground. The sink is the heating distribution system used for the upgraded heat.

The key to ensuring a high efficiency heat pump system is to collect the heat energy at the highest possible temperature and to distribute the heat at the lowest possible temperature. This is why in winter, a ground source heat pump supplying an efficient underfloor heating system can be a very efficient heating system, as the ground is warmer than the air and the underfloor heating system has a lower distribution temperature than radiators. To make radiators effective on a heat pump circuit, the radiators need to be oversized compared to standard radiator sizes to realize the lower distribution temperatures.

> **Heat Source** In relation to heat pumps, the heat source is heat which can be extracted for use within a heat pump. Heat sources for heat pumps include air, ground and water.

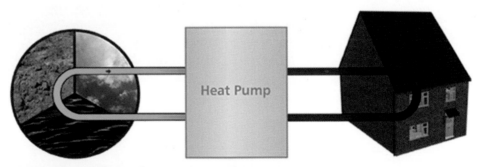

Illustration of how a heat pump works

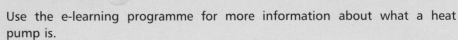

E-LEARNING

Use the e-learning programme for more information about what a heat pump is.

ACTIVITY 1

Heat pumps normally run on electricity. A traditional combustion boiler runs on gas, oil or solid fuel. What could you do to run a heat pump and a combustion boiler on 100 per cent renewable energy instead of fossil fuel?

Selecting a heat pump

UK AND INTERNATIONAL STANDARDS

Heat pumps are suitable for most types of properties, but older properties tend to suffer much higher heat loss than those built to modern standards. Before considering the installation of a heat pump, the insulation and air tightness of the building should be measured for energy efficiency against the latest building regulations, such as Part L in England and Wales.

Similar energy efficiency building regulations now apply in most European countries. Unless the building is brought up to current standards, the efficiency of the heat pump will be affected. Money spent on insulation, draught-proofing and modern, well-fitting glazing can be recovered in the cost savings of running a heat pump.

Houses can lose a lot of heat through the windows and roof if they are not properly insulated

The heat loss of the building must be calculated before selecting a heat pump. The heat loss of the building is calculated by evaluating the heat lost through the walls, floor and roof together. This will ensure the correct size and type of heat pump is selected for the required heat output.

Heat pumps are usually sized to meet the maximum heat load requirement of the building. An undersized or oversized heat pump can affect the cost, efficiency and effectiveness of the system.

MCS STANDARDS

A free-to-use spreadsheet heat loss calculator is available on the British MCS website. This heat loss calculator can be used with the appropriate national weather data to calculate the heat loss in any European country.

Heat loss must be calculated before selecting a heat pump

Air Source Heat Pump (ASHP) A system which extracts heat from (or expels heat to) the outside air to upgrade the heat in a heat pump in order to heat (or cool) a building. Most ASHPs are heat only.

Ground Source Heat Pump (GSHP) A system that uses heat from the ground to heat (or cool) a building. It uses the Earth as a heat source in the winter or a heat sink (in cooling mode) in the summer.

ACTIVITY 2

A swimming pool is to be heated during the summer months only using a heat pump. What would be the best option for heating this system, an **air source heat pump** or a **ground source heat pump**? This heat pump system is fitted on a summer only holiday home and another heat pump system is to be provided for hot water heating. Which heat pump circuit will be more efficient, the swimming pool system or the hot water system? Please explain your answers.

THE EFFICIENCY OF HEAT PUMPS

Calculating the coefficient of performance

The efficiency of a heat pump is measured by its **coefficient of performance (CoP)**. This is calculated by dividing the heat output of the pump in **kilowatts (kW)** by the electrical input in kilowatts. For example, a pump that delivers 4kW of heat output for every 1kW of electricity used to power the process has a CoP of 4. This is sometimes expressed as a ratio, i.e. 1:4 or as a percentage, i.e. 400 per cent.

FUNCTIONAL SKILLS

The coefficient of performance is calculated using the following equation:

$$CoP = \frac{Heat\ Output(kW)}{Electricity\ Input(kW)}$$

For example:

$$CoP = \frac{4kW}{1kW} = 4(1{:}4\ \ or\ \ 400\%)$$

> **Coefficient of Performance (CoP)** A measure of the heat pump's efficiency. The CoP is calculated by dividing the heat output of the pump in kilowatts by the electrical input in kilowatts. CoPs can be expressed as a number, ratio or percentage. (e.g. '4', '4:1' or '400 per cent').

> **Kilowatts (kW)** The kilowatt is equal to one thousand watts. This unit is used to express the output power of engines, the power consumption of tools and machines, the heating and cooling power used and any other forms of power.

Heat Output (kW)
e.g. 4kW

Electricity Input (kW)
e.g. 1kW

The efficiency of a heat pump is calculated by dividing the output by the input

CoPs typically vary between approximately 2 and 5.4, for different makes and models of heat pumps in different applications. So that different heat pumps can be compared, measurements are often taken under standard test conditions. For ground source heat pumps, CoP is usually quoted on a source water temperature of 0°C and an output water temperature of 35°C. As you can see from the chart below, varying from these figures can affect the CoP value dramatically.

Air source heat pumps are quoted at a wider variety of values such as 2 or 7°C source temperature and 35°C sink temperature.

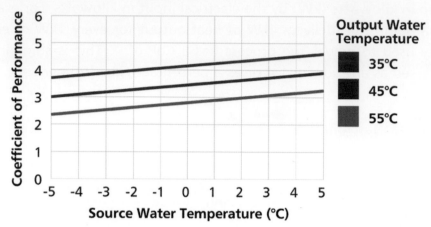

The coefficient of performance is affected by the source water temperature and the output water temperature

ACTIVITY 3

A property has an electricity bill of 3650 kWh per annum at a cost of £545.00 per annum before tax. Your customer who owns the property asks you to quote them for a ground source heat pump (GSHP) system including their predicted spend on space heating and hot water heating. You know that the split between space heating and hot water is 75:25 and the house requires 12 000 kWh per annum. How much will the predicted spend be? The GSHP will have a CoP of 3.7 in space heating mode and 2.8 in hot water mode.

Calculating the seasonal performance factor

The seasonal performance factor (SPF) is an important value because it measures the performance of a heat pump over a full season of use. It can be considered as the installed CoP and is calculated by dividing the total actual amount of heat produced by the heat pump over a season by the total actual electricity consumption of the heat pump over the same season.

Factors that can affect the SPF are the annual heating demand, the peak load of the property, the input source temperature and the required heating temperature, the energy efficiency of the heat pump components, the operating characteristics of the heat pump, the sizing of the heat pump in relation to demand, the heat pump or building control system and any other variables that affect the heating circuit.

FUNCTIONAL SKILLS

The seasonal performance factor (SPF) is calculated by the equation:

SPF = Sum of Heat Delivered (kWh) / Sum of Electricity Consumed (kWh)

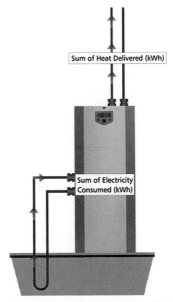

The seasonal performance factor is calculated by dividing the sum of heat delivered by the sum of electricity consumed

Flow Temperature
The temperature of the water leaving a source of heat, in this case, the heat pump.

ACTIVITY 4

The SPF on a property with an air source heat pump (ASHP) is currently 2.1. The heat pump is used to provide space heating only. The radiators are over-sized by a factor of 1.6 which, according to the heat emitter guide, indicates that the **flow temperature** of the ASHP is 60°C. There is limited draught proofing, single glazing and unfilled cavity walls in the property. What advice would you give the householder to improve the SPF and what final SPF might be realized with extensive changes to the property? The heat emitter guide can help with this question, especially studying the temperature star rating and its possible improvement. Write some general notes here.

CHECK YOUR KNOWLEDGE

1. **How is the seasonal performance factor (SPF) of a heat pump calculated?**

    Complete the equation shown by using these 3 options:

    ☐ a. Sum of electricity consumption

    ☐ b. SPF

    ☐ c. Sum of delivered heat

    $$ = \underline{\hspace{5cm}} $$

2. **Which part of the current Building Regulations relates to energy efficiency?**

    Circle the correct letter from the options shown:

A	B	C	D	E	F	G	H	I	J	K	L	M
N	O	P	Q	R	S	T	U	V	W	X	Y	Z

Chapter 2

HEALTH AND SAFETY

LEARNING OBJECTIVES

By the end of this chapter you will be able to:

List the key items of Personal Protective Equipment

- Identify common safety and hazard signs

- Describe the effects of hazardous substances

- Describe groundwork hazards

- Identify the key elements of a first aid kit

- Know how to carry out a risk assessment

- Match fire extinguishers to different types of fire

- Select CSCS safety cards for different types of people

- Describe all aspects of electrical safety

HEALTH AND SAFETY AT WORK ACT

HASAWA (1974)

The Health and Safety at Work Act (1974) provides the legal framework to promote, stimulate and encourage high standards of health and safety in the workplace.

It protects employees at work and the public in the workplace and gives everyone the responsibility for their own health and safety and the health and safety of others who could be affected.

Everyone has a duty to comply with the Act, including employers, employees, trainees, the self-employed, manufacturers and suppliers.

Before going any further though, it is essential that you have an understanding of the Health and Safety at Work Act, 1974.

If negligence can be proved under HASAWA, both employers and employees can face a £5000 fine from a Magistrate's Court and unlimited fines and imprisonment from a Crown Court.

HASAWA (1974)

Employer responsibilities

The Act places a general duty on employers to ensure 'so far as is reasonably practicable, the health, safety and welfare at work of all their employees'.

Employers must provide and maintain safety equipment and safe systems of work. This includes, among other things, ensuring materials are properly stored, handled, used and transported, providing information, training, instruction and supervision and ensuring staff are aware of manufacturers' and suppliers' instructions for equipment. Employers must also look after the health and safety of others, for example the public, and talk to safety representatives.

Employers are forbidden to charge employees for any measures which are required for health and safety, for example, personal protective equipment (PPE).

Employer

Employee responsibilities

Employees must comply with the Act and look after their own health and safety as well as the health and safety of others. They must co-operate with their employers and not interfere with anything provided in the interest of health and safety.

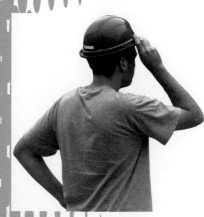

Employee

Personal protective equipment (PPE)

Personal protective equipment (PPE) covers a range of different items of clothing or equipment – such as gloves or safety helmets – that are used at work to avoid harm or injury.

Employers have a legal duty to identify any risks involved with a particular job and therefore provide any items of PPE that may be needed – but you still need to know the basics so you can make sure you get what you need.

Ideally risks to health and safety will be eliminated from the workplace before they occur, so PPE should be the last line of defence. However, if risks remain, then PPE must be used and provided to you free of charge.

Worker wearing PPE while at work

Common PPE items

You need to be familiar with several key items of PPE. You may not need to use them on every job, but you still need to know when they are required and how they should be used.

Safety footwear

For many jobs you will need to wear steel-toed boots with intersoles – these are thin pads in the boots or shoes that absorb surface shock – to protect your feet from injury.

If working in wet conditions, rubber boots should be worn – which must also have intersoles and steel toecaps.

safety footwear

Overalls

Overalls (also known as boiler suits) are ideal for many jobs as they provide cover for your entire body. However, you must never wear overalls made from terylene, nylon or similar materials as these catch fire easily.

Overalls

Ear muffs/ear plugs

If you are exposed to high noise levels, you must protect your hearing with ear muffs or ear plugs.

Ear muffs must always be properly fitted so that the ear is completely covered, otherwise you will not be fully protected.

Ear plugs fit inside the ear and are often disposable. They offer less protection than ear muffs.

Ear muffs

Respirators

Many different types of respirator are available, but filter masks are the most common. These are rubber face masks that fit over the mouth and nose, containing a filter canister through which the wearer breathes.

Filter masks stop dust but do not provide protection against harmful gases or vapours. If you are working around harmful gases or vapours, you must use a canister respirator. Filter canisters must be changed regularly.

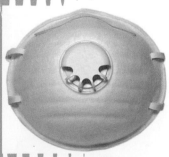

Respirator

Safety gloves

Safety gloves protect your hands from injury. Take care to choose the right kind, as different types exist, for example when working with heat or chemicals. Check the gloves' application data, if available.

Do not wear gloves when using machinery – e.g. drills – as this is dangerous.

Safety Gloves

Safety goggles

You must wear safety goggles when required, for example when welding, working in dusty conditions, or when flying chippings will be produced.

Always check that you have the right type of goggles, as different lenses offer different levels of protection.

If goggles have dirty lenses, clean them before use: never obscure your vision.

Safety goggles

Safety hats/helmets

Hard hats should be worn on a building site whenever there is a risk of falling or flying objects. You must always check that your hat fits properly. Always wear your hat with the peak facing forward, because the peak lip is designed to protect your eyes. When working indoors, a 'bump hat' can be worn instead.

Hard hats sometimes come with an expiry date and should not be used after this date as it may have deteriorated and may not provide the necessary protection.

Safety helmet

E-LEARNING

Use the e-learning programme to learn more about the different types of PPE.

SAFETY SIGNS

Work sites make extensive use of safety signs to warn of a number of different risks and hazards. You will learn what the different signs mean, but before you do it is important to understand what is meant by a 'hazard' and 'risk'– as they are not the same thing.

Safety sign

Hazards

A hazard is something that can potentially damage your health. The negative effects of a hazard may be relatively minor, such as making your eyes water, or they may be much more serious, such as suffocation and in some cases, they can be fatal.

Hazards may be difficult to detect and might not affect you immediately, such as long term contact with asbestos which can cause cancer.

Risk

Risk refers to how likely it is that a hazard will actually damage your health. The better controlled a hazard is, and the more rigorous people are about using PPE properly, the lower the risk will be.

Identifying safety signs

To be able to work safely you must understand what the basic safety signs mean.

Most signs usually have accompanying textual information to further explain the sign.

'Prohibition' signs, which are circular with red crosses through them, tell you not to do something.

No smoking   *No lit matches*   *Do not extinguish with water*   *No pedestrians*

'Mandatory' signs, which are blue circles with white symbols, tell you what you must do.

Must wear safety goggles   *Must wear safety gloves*   *Must wear ear muffs*   *Must wear hard hat*

'Information' signs, which are green squares with white symbols, are used to communicate safety information.

Emergency phone

Emergency exit direction

First aid

Hazard signs

Hazardous substances are identified by four main warning signs.

'Warning' signs, which are yellow triangles with black symbols, identify a particular hazard or danger.

Dangerous chemicals

Danger acid

Danger

Irritant

E-LEARNING

Use the e-learning programme to learn more about the signs and what each one means.

CONTROL OF SUBSTANCES HAZARDOUS TO HEALTH (COSHH)

Classifying hazardous substances

Hazardous substances can be broken down into four main categories:

- Toxic/very toxic substances
- Corrosive substances
- Harmful substances
- Irritants

Being able to identify these in a practical setting will help to alert you to possible sources of danger so that you can take appropriate protective action.

Toxic/very toxic substances

Toxic/very toxic substances, such as bleach, can cause death or serious damage when inhaled, swallowed or absorbed by the skin.

Corrosive substances

Corrosive substances, such as sulphuric acid, can destroy parts of the body if they come into direct contact with them.

Harmful substances

Harmful substances, such as lead, can cause death or serious damage when inhaled, swallowed or absorbed by the skin.

Irritants

Irritants, such as soft solder flux, can cause inflammation or swelling if they come into contact with the skin, eyes, nose or mouth. Such effects may be felt immediately, or they occur after extended or repeated contact.

E-LEARNING

Use the e-learning programme to learn more about hazardous substances.

UK AND INTERNATIONAL STANDARDS

Under European law, substances are defined as being officially hazardous to health if they are listed as 'dangerous to supply'. They must also be defined as being very toxic, toxic, corrosive, harmful, or irritants. One practical result of this is that when sold commercially, the packaging for such substances must be clearly marked and labelled.

Substances hazardous to health

Corrosive material

Substances marked as being 'corrosive' could cause permanent damage if they come into direct contact with any part of your body. For example, sulphuric acid will burn your skin away and cause breathing problems.

Flammable material

Flammable materials must be kept away from naked flames and you must not smoke when near them. Common flammable materials include your own clothes, some modern hair products and oily rags that may have been left lying around.

Explosive material

Explosive materials must be handled and stored with particular care, as they potentially present an extreme hazard.

Toxic material

Some toxic materials, such as gas in a confined area, can harm you even if you do not come into direct contact with them, so ensure you handle them with care. If necessary, seek advice as to whether PPE is required when they are in use.

The effects of hazardous substances

In order to be able to adequately follow safety procedures and use PPE correctly, it is important to understand exactly how harmful substances can affect your body.

Skin and eyes

Some substances will cause damage, such as burns, if they come into direct contact with the skin or eyes. Contact may also lead to less

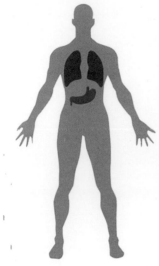

Hazardous substances can damage skin, eyes, lungs and stomach

serious problems, such as increased skin sensitivity. Harmful substances, such as solvents, can also enter the bloodstream by being absorbed through the skin, or through cuts and abrasions.

Lungs

Harmful substances can cause damage when inhaled in one of two ways. Either they can cause direct damage to the lungs themselves e.g. asbestos poisoning, or they can enter the bloodstream and so affect other organs.

Stomach

Harmful substances, such as lead, can reach the stomach in numerous ways if basic hygiene is not observed or if gloves are not worn. Eating, drinking, smoking and biting your nails after coming into contact with the substance can all be responsible for this.

E-LEARNING

Use the e-learning programme to learn more about the effects of hazardous substances.

Groundwork hazards

Boreholes A narrow shaft drilled vertically into the ground. These can be used to accommodate vertical collector pipes with a ground source heat pump. Boreholes for ground source heat pumps can range from 5m to 150m deep.

When planning any groundworks and excavations such as trenches, **boreholes**, pits and tunnels, it is important to have measures in place to prevent accidents.

Potential hazards include:

- collapse of the sides of the trench
- materials falling onto anyone working in the trench
- people or vehicles falling into the trench
- people being struck by plant
- undermining nearby structures
- coming into contact with any underground services
- the access to the excavation
- any fumes that may be released.

Care should be taken when preparing any groundworks

Collapse of the sides

If an excavation is not correctly supported there is a risk that the sides of the excavation may collapse. The type and moisture content of soil, the ground water conditions, the type and depth of the excavation all affect the amount of support required to prevent a collapse.

When working in an excavation, work should be properly supervised and inspected by a competent person. Inspections should be carried out before a work shift, after an event which affected the stability of the excavation and after accidental falling of materials into the excavation.

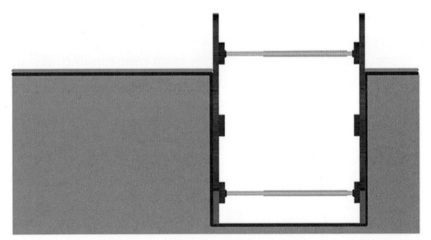

Sides of excavation may collapse if not correctly supported

People falling into the excavation

Guarding should be in place to prevent people from falling into excavations. There are a number of ways to guard excavations. These include guard rails and toe boards, fabricated guard rail assemblies or using the support system itself. This could include making the trench sheets longer than the depth of the trench or by using trench box extensions. In order to prevent people from approaching excavations in public areas, suitable fencing should be in place.

Ladders should only be used for accessing the excavation if it is deemed safe to do so. The ladder itself must be suitable, secured and extend at least 1m above the landing place.

Guards should be put in place to prevent people from falling into excavations

Vehicles falling into the excavation

Vehicles travelling too close to the edges of an excavation could potentially overload the sides of the excavation, causing it to collapse. To prevent this, suitably painted or marked baulks or barriers should be in place to prevent vehicles from travelling near the edges. Where vehicles are required to unload materials into excavations, stop blocks should be properly secured to prevent the vehicles from over-running. The sides of the excavation should also have extra support to prevent the edges from breaking or the sides collapsing under the weight of the vehicle.

Precaution measures should be in place to prevent vehicles from falling into excavations

Materials falling into the excavation

Vehicles, machinery, excavated spoils and any other materials should not be parked or placed near the edges of the excavations as the extra weight can overload the sides of the excavation causing it to collapse. Petrol, diesel and other hazardous fumes should also be kept away from the excavation unless it is ventilated. Finally, edges should be protected by toe-boards or trench sheets to protect workers against falling materials, and head protection should be worn when working in an excavation.

Materials should not be placed near the edge of excavations

Undermining nearby structures

Undermining nearby structures such as garden or boundary walls which have shallow foundations can cause these structures to collapse. Other structures such as scaffold footings, underground services or foundations of nearby buildings should also not be undermined. It should be decided whether extra support is required before digging starts. In some cases, surveys of foundations and advice from structural engineers should be sought.

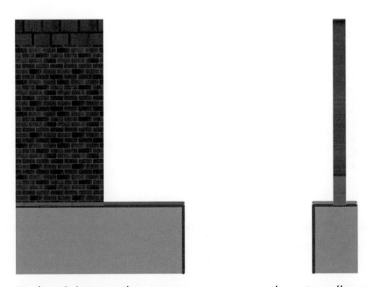

Undermining nearby structures can cause them to collapse

Underground services

Underground services such as electricity cables or gas pipes can cause serious accidents if they are damaged during excavations. Excavation work should not begin until a locator guided by service plans is used to accurately locate all underground services. Once underground services are located, digging near to these services should be done by hand and extra care should be taken to spot any damage made to the services. Always treat services as live, and therefore dangerous, unless it has been specifically stated otherwise. In the event of damaged gas pipes, ban smoking and naked flames, evacuate if necessary and erect suitable warning signs.

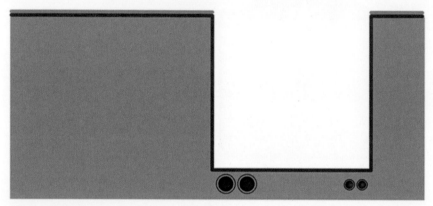

Look out for underground services such as electricity cables or gas pipes

Overhead services

Overhead services such as overhead power lines can also cause serious accidents, such as burns and electrocution, if raised tipper trucks or excavators come into contact with them.

For both underground and overhead services, an emergency plan must be in place to notify service owners if services are damaged. Finally, plans should be updated once new services are laid.

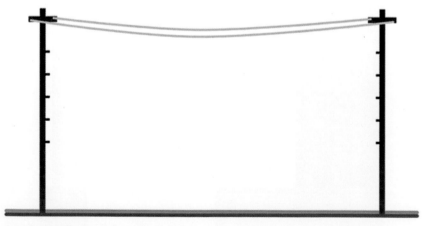

Look out for overhead services

Surface water inflow

The permeability of the ground should be taken into consideration when designing the supports for the sides of the excavation. This is particularly important for areas close to lakes, rivers or the sea. Water that enters the excavation needs to be channelled to sumps so that it can be pumped out. Another consideration for designers is the effect on the stability of the excavation when the water is being pumped out. Alternative de-watering techniques include ground freezing and grout injection.

Precautions must be taken when working in areas close to lakes, rivers or the sea

E-LEARNING

Use the e-learning programme to learn more about groundwork hazards.

BASIC FIRST AID

However well prepared people are, accidents will happen. Employers have a legal duty to ensure that first-aid kits are in place, containing sufficient supplies for all workers onsite.

For smaller work sites, clearly marked first-aid boxes must be placed under the control of a named individual. For larger sites of over 50 people, there must also be at least one person with first-aid training.

First Aid kit

MANUAL HANDLING

Manual handling is any activity of transporting or supporting a load (i.e. an object) by hand or with bodily force and includes lifting, carrying, lowering, pushing and pulling.

Lifting

Lifting can be broken down into five steps:

1. Stop and think – plan the lift and assess the risks.
2. Take up a good posture to start the lift and keep your back naturally straight.
3. Take a firm grip on the load, no twisting or over-reaching.
4. Start to lift using power of the legs to make a smooth and controlled action. If the load is too heavy do not continue.
5. Keep the load under control, heaviest side closest to the body, and beware of liquids and uneven loads.

1. Plan the lift and assess the risks

2. Take up a good posture, keeping your back straight

3. Take a firm grip on the load, avoid twisting or over-reaching

4. Start to lift using the power of your legs, if the load is too heavy do not continue

5. Keep load under control

Carrying

Carrying the load is usually the main purpose of manual handling. When carrying the load move your feet steadily and slowly, look ahead (not at the load) and avoid twisting your body.

- Keep the load close to your body to maintain control.
- Be vigilant and aware of your surroundings – especially be alert to possible danger, in addition to original assessment of risks. Be ready for the unexpected.
- If feeling tired or strained – stop. Don't overdo it.

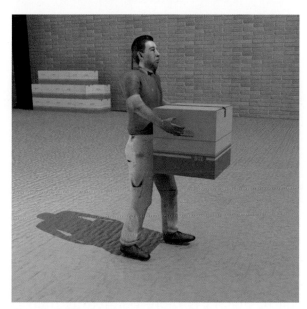

Carrying the load

Pushing

For loads that are too large or too heavy to be carried by hand, a manual handling aid such as a trolley, wheelbarrow or scissor lift should be used.

- Always check the manual handling aid is in good working order.
- Assess the whole route for hazards, especially slopes up or down and uneven surfaces.
- Use safe lifting techniques to place the load onto the manual handling aid.
- Take a firm grip on the handles and lean forward slightly, keeping a natural straight back.
- Use the power of the legs to push in an even and controlled way, with a slow walking pace.
- Once moving, keep a steady pace and keep feet away from the load.

- When you arrive at the destination slow down smoothly (no jerking movements) and unload in the same way it was loaded.

- Always return the manual handling aid to its storage place to avoid turning it into a hazard, or preventing someone finding it when needed. If the aid cannot be found they may be tempted to use unsafe manual handling methods.

Pushing

Pulling

When you have a load on a trolley, or other manual handling aid, you can push or pull it. Pulling is more dangerous than pushing as the load can 'run over' you. If you lose control when pushing, the load will move away from you, if however you were pulling, the load will run towards you. Pulling also puts more strain on the back and it is harder to control the load.

- Always check the manual handling aid is in good working order.

- Assess the whole route for hazards, especially slopes up or down and uneven surfaces.

- Use safe lifting techniques to load the trolley.

- Take a firm grip on the handles and lean backward slightly, keeping a natural straight back.

- Maintain a steady pace, a slow walking pace, and keep feet away from the load.

- When you arrive at the destination slow down smoothly (no jerking movements) and unload in the same way it was loaded.

- If uncomfortable at any time stop and reassess the situation. You may need to lighten the load or get help.

Pulling

Good manual handling practice

Carrying the load is the main purpose of manual handling. When carrying a load move your feet steadily and slowly, look ahead and not at the load and avoid twisting your body. Keep the load close to your body to maintain control – especially if the load is difficult to hold, e.g. a container of liquid or odd shaped object. Be vigilant and aware of your surroundings and especially be alert to possible danger. Your original assessment of risks was taken at a particular time – things may have changed so be ready for the unexpected. Remember, if you are feeling tired or strained – stop. Don't overdo it.

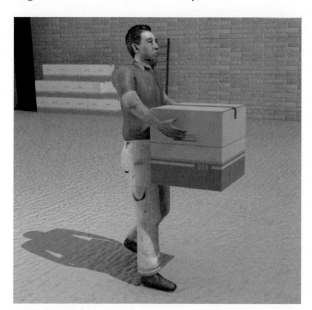

Carrying the load

ASSESSING HEALTH AND SAFETY RISKS IN THE WORKPLACE

Many accidents in the workplace can be avoided if appropriate health and safety risk assessments are performed. This is why it's very important to assess risks in the workplace, in order to protect not only the people working there, but also members of the public.

There is a legal requirement to carry out regular risk assessments, and although the law does not expect all risks to be eliminated, people must be protected as far as it is reasonably practicable to do so.

Health & safety risks in the workplace

Risk assessment

Risk assessment is the process of identifying hazards in the workplace and then deciding who might be harmed and how.

A hazard is anything that might cause harm, for example electricity, chemicals, working from ladders, or an open drawer.

There are five steps in the risk assessment process. Once this process has been completed, suitable precautions can be put in place to reduce the risk of harm or to make the potential harm less serious.

The five steps of risk assessment are:

- Identifying the hazards
- Identifying who could be affected

- Evaluating the risks
- Recording findings and implementing them
- Reviewing risk assessment.

Ladders can be considered as a hazard

Step 1 – Identify the hazards

The first step is to identify what the hazards are. When you work in a place every day, it's easy to overlook potential hazards, so there are ways to make sure you identify those that matter.

Walking around to look for things that might reasonably be expected to cause harm is a good starting point. Also, ask employees or their representatives, as they might be aware of things that are not immediately obvious to you.

Publications and practical guidance are available from a variety of sources, including the Health and Safety Executive (HSE) and relevant trade associations, as well as manufacturers' instructions. These all provide information on where hazards can occur, their harmful effects and how to control them.

Accident and ill-health records are another source of information which can often help to identify the less obvious hazards. Don't forget the long-term health hazards, for example, those that can occur following prolonged exposure to high levels of noise, or harmful substances.

Trailing cables

Spills can cause people to slip

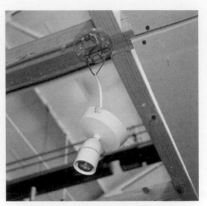

Faulty/damaged electrical fittings can be dangerous

Items left lying about could cause accidents

Incorrect manual handling

Exposure to excessive noise may cause hearing loss

Here are some potential hazards:

- Trailing cables might cause people to trip
- Spills might cause people to slip and also be toxic
- Faulty or damaged electrical fittings might cause electric shocks and possibly a fire
- Items left lying about might cause people to trip or block access
- Lifting heavy loads might cause back injuries
- Long-term exposure to excessive noise might cause loss of hearing

Step 2 – Who might be harmed and how

The second step in risk assessment starts with deciding who might be harmed by each of the identified hazards.

To do this, it's best to identify groups of people rather than listing people by name, as this will help later on when it comes to identifying the best ways to manage the risk for each group.

It's also necessary to identify how the different groups of people might be harmed, for example, what type of injury or ill-health might occur.

Here are some groups of people who have different requirements:

- Customers or members of the public who are visiting the workplace
- Contractors who might not be in the workplace all the time
- Maintenance workers visiting the site
- People with disabilities
- Store workers responsible for moving heavy boxes
- Pregnant women

Customers

Contractors

Maintenance workers

People with disabilities

Store workers

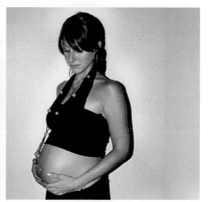
Pregnant women

Step 3 – Evaluate the risks and decide on precautions

Having spotted the hazards and worked out which groups of people might be affected by them and then how, the next step is to decide what needs to be done to protect people from harm.

The law requires that everything 'reasonably practicable' is done and the easiest way to find out if this is the case is to compare what is already being done with good practice.

Evaluating the risks and deciding on what precautions to take

The first thing to consider is whether or not the hazard can be eliminated. If not, there are certain actions that should be applied in the order shown, to control the risk and making harm less likely.

Improving health and safety doesn't need to cost a lot and failure to take simple precautions could cost a lot more if an accident occurs. Before introducing new precautions, always check that these are reasonable and do not introduce any new hazards.

No trespassing   *Must wear hard hat*   *Danger*   *First aid*

Here are some different actions that could be applied:

- Switch to using a less hazardous method
- Switch to using a less hazardous chemical
- Prevent access by guarding the hazard

- Provide lifting equipment
- Provide PPE – protective clothing, footwear, goggles, etc.
- Provide first aid and washing facilities for removal of contamination

Step 4 – Record findings and implement them

After you've spotted the hazards, worked out which groups of people might be affected by them and how, then decided what needs to be done to protect people from harm, it's important to keep a record of what's been done.

In fact, for businesses with five or more employees, the results of the risk assessment must be written down and actions recorded as they are implemented.

Smaller businesses will also find it useful to have a written record of their risk assessment so it can be reviewed at a later date.

You should keep the written records of the risk assessment as simple as possible and share the document with all employees.

If many improvements need to be made, it is best to draw up an action plan to deal with the most important ones first rather than trying to do everything at once.

Recording findings

A good action plan will often include implementing a few cheap or easy improvements which can be done quickly, perhaps as a temporary solution, until more reliable controls can be put in place.

The plan of action might also include long-term solutions for those risks which are most likely to cause accidents or ill-health, or which have the worst potential consequences, as well as the arrangements which are made for training employees on how the remaining risks will be controlled.

Finally, the action plan should include who has responsibility for the various actions and the recommended deadline for implementing these actions, as well as how often checks will be made to make sure that control measures stay in place.

Plan of action

Step 5 – Review risk assessment and update if necessary

Having identified the hazards, who might be harmed by them and how, what the risks are and the controls that are needed, and kept a record of what has been done, the fifth step in the risk assessment process is to review the risk controls and to update them as necessary.

It is a good idea to review the risk assessment on an ongoing basis, by thinking about risk assessment when changes are being planned, as well as conducting a formal, annual review.

Regular reviews must be carried out

Reviewing the risk assessment regularly will mean that controls can be amended each time new hazards are introduced, for example when there are significant changes in the workplace, or with the introduction of new equipment, substances or procedures, when problems are spotted by employees, or when accidents or near misses occur.

By carrying out regular reviews you can ensure that risk controls are always up to date and improving and not sliding back.

Responsibilities

Both employers and employees have responsibilities for health and safety in the workplace.

Employers are responsible for ensuring risk assessments are carried out on a regular basis. The process does not need to be a complicated one nor do you need a health and safety expert to do one.

Employees have a responsibility to co-operate with their employer's efforts to improve health and safety by complying with the controls which are in place and by looking out for each other.

Company name:		Date of risk assessment:		
Step 1 What are the hazards?	**Step 2** Who might be harmed and how?	**Step 3** What are you already doing?	What further action is necessary?	**Step 4** How will you put the assessment into action?
Spot hazards by: ■ walking around your workplace; ■ asking your employees what they think; ■ visiting the Your industry areas of the HSE website or calling HSE Infoline; ■ calling the Workplace Health Connect Adviceline or visiting their website; ■ checking manufacturers' instructions; ■ contacting your trade association. Don't forget long-term health hazards.	Identify groups of people. Remember: ■ some workers have particular needs; ■ people who may not be in the workplace all the time; ■ members of the public; ■ if you share your workplace think about how your work affects others present. Say how the hazard could cause harm.	List what is already in place to reduce the likelihood of harm or make any harm less serious.	You need to make sure that you have reduced risks 'so far as is reasonably practicable'. An easy way of doing this is to compare what you are already doing with good practice. If there is a difference, list what needs to be done.	Remember to prioritise. Deal with those hazards that are high-risk and have serious consequences first. Action Action Done by whom by when

Risk Assessment document

FIRE PROTECTION

Classes of fire

There are four classes of fire:

- Class A: Solids e.g. wood, paper, textiles
- Class B: Flammable liquids e.g. oil, petrol, paint
- Class C: Flammable gases e.g. acetylene, propane, butane
- Class D: Metals e.g. magnesium, aluminium, sodium

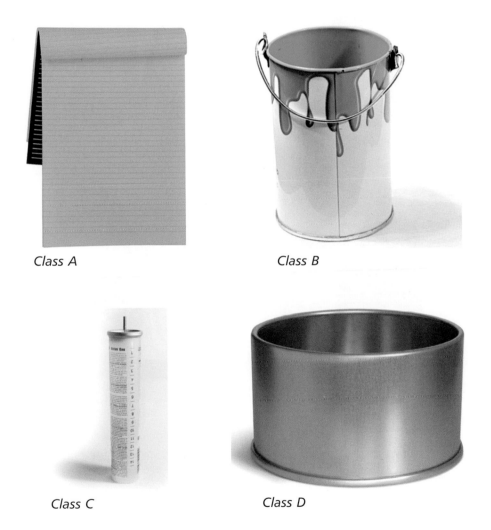

Class A

Class B

Class C

Class D

The type of material that is on fire tells you which type of fire extinguisher to use.

As electrical fires do not fall into any particular 'class' of fire, if an electric spark ignites, for example, paper, you would use a Class A extinguisher.

Types of fire extinguisher

There are four different types of fire extinguisher, which are shown below.

Water fire extinguisher

Foam fire extinguisher

Powder fire extinguisher

Carbon dioxide fire extinguisher

All fire extinguishers are now red and are labelled to identify which type is which. It is important to select the correct one for the class of fire otherwise it could have serious consequences.

- A water fire extinguisher is suitable for class A fires
- A carbon dioxide fire extinguisher is suitable for class B, C and D fires

- A foam fire extinguisher is suitable for class B and D fires
- A powder fire extinguisher is suitable for all classes of fire (A, B, C and D)

CSCS SITE SAFETY CARDS

There are nine different types of site safety card in the Construction Skills Certification Scheme (CSCS).

CSCS Safety Cards

Each card is issued to people in relation to their relevant experience and qualifications, and requires completion of an appropriate health and safety course or test.

Red card

> **Red Card – Trainee (Craft and Operative):** registered for NVQ or SVQ (or Construction Award) and not yet reached Level 2 or 3.
>
> **Red Card – Trainee (Technical, Supervisory and Management):** registered with a further/higher education college for a nationally-recognized, construction-related qualification or satisfactorily completed such a course.

Green card

> **Green Card – Construction Site Operative:** who carries out basic site skills with Level 1 NVQ or employer's recommendation using industry accreditation.

Blue card

> **Blue Card – Experienced Worker:** with at least one year's on-the-job experience in last three years who missed opportunity for industry accreditation. Card is valid for one year whilst achieving Level 2 or higher NVQ/SVQ but is not renewable.
>
> **Blue Card – Skilled Worker:** with Level 2 or higher NVQ/SVQ or completed employer sponsored apprenticeship or completed City & Guilds of London Institute Craft Certificate.
>
> **Blue Card – Experienced Manager:** with at least one year's on-the-job experience in last three years. Card is valid for three years whilst achieving Level 3 or higher NVQ/SVQ but is not renewable.

Gold card

> **Gold Card – Skilled Worker:** with Level 3 NVQ/SVQ or approved indentured apprenticeship or employer sponsored apprenticeship and completed City and Guilds of London Institute Advanced Craft Certificate.
>
> **Gold Card – Supervisor:** with Level 3 NVQ/SVQ in supervisory occupation or industry accredited.

Platinum card

> **Platinum Card – Manager:** with industry accreditation or Level 4 NVQ/SVQ.

Black card

> **Black Card – Senior Manager:** with Level 5 NVQ/SVQ.

Yellow card

> **Yellow Card – Professionally Qualified Person:** consultants who are chartered members of approved institutions (e.g. architects, surveyors and engineers) with health and safety responsibilities and on site no more than 30 days in six month period.
>
> **Yellow Card – Regular Visitor:** with no specific construction skills who often visits a construction site. Makes site access easier as the holder would have passed a health and safety test before the card was issued.

White card

> **White Card – Construction Related Occupation:** for those occupations not covered by other cards.

ELECTRICAL SAFETY

Looking after and maintaining electrical equipment

All electrical equipment should be looked after and maintained in a safe condition.

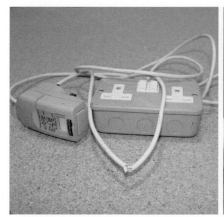
Extension leads that are moved a great deal are prone to damage

Plug cables should be firmly clamped

Always use proper connectors to join cables

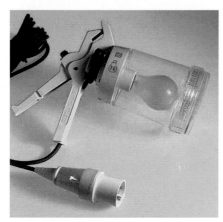
Equipment that can easily be damaged must be protected

Suspect or faulty electric equipment must be labelled "Do not use"

Some equipment is not suitable for wet or harsh environments

- Continuously moving extension leads makes them prone to being damaged. If the cable, plug or socket are damaged they should be replaced.

- Plug cables should always be firmly clamped. The outer sheath of flexible cables must always be firmly clamped to stop the wires (particularly the earth) from pulling out of the terminals.

- Always use proper connectors to join cables. Cables should always be joined with proper connectors or cable couplers, not with strip connectors or insulating tape.

- Lamps and equipment that can easily be damaged must be protected to prevent risk of electric shock.

- Suspect or faulty electrical equipment must be labelled 'DO NOT USE' and kept secure until it can be examined by a competent person.

- Equipment unsuited for use in a wet or harsh environment can easily become live and also make the surroundings live.

Visual inspection

Many faults with electrically operated power tools can be found by visual inspection, and by following a simple process before using the equipment you can minimize most electrical risks, as shown here:

- Switch off and unplug

- Check plug is correctly wired

- Check fuse is correctly rated by checking equipment rating plate or instruction book

- Check plug is not damaged, cable is properly secured and no internal wires are visible

- Check cable is not damaged and has not been repaired with insulating tape or unsuitable connector

- Check outer cover of equipment is not damaged which might give rise to electrical or mechanical hazards

- Check equipment for burn marks or staining that might suggest equipment is overheating.

Switch off

Unplug the equipment

Check plug is correctly wired

Check fuse is correctly rated

Check plug is not damaged

Check cable is not damaged

Check outer cover of equipment is not damaged

Check equipment for burn marks

E-LEARNING

Use the e-learning programme to see a demonstration of a visual inspection.

Electrical site safety

UK AND INTERNATIONAL STANDARDS

When it comes to the safe isolation of electrical supplies and energizing electrical installations, it is also important to comply with statutory health and safety requirements, as laid down by the Electricity at Work Regulations.

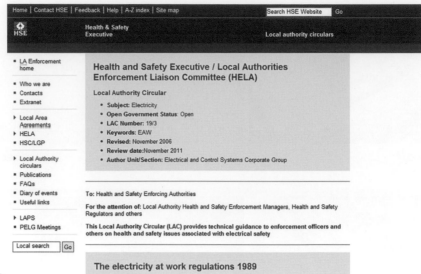

HSE Website

Safe isolation of electrical supplies

In order to avoid fatal accidents which can occur during the proving of isolation, you should follow the recognized procedure:

- Identify source of supply
- Identify type of supply
- Isolate
- Secure the isolation
- Test the equipment/system is dead
- Begin work.

Identify the source of supply

Identify the source of supply

Isolate

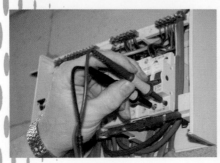

Secure the isolation

Test the equipment/system is dead

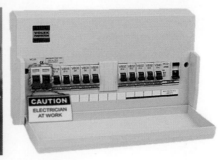

Begin work

E-LEARNING

Use the e-learning programme to see a demonstration of safe isolation.

Energizing electrical installations

A number of deaths and major injuries have occurred when electrical circuits have been energized at the request of building designers, clients, contractors or finishing trades before the electrical installation was complete. It is not considered 'reasonable to work live' solely on the grounds of inconvenience, lost time, or cost.

Electrical contractors are only able to energize circuits when it is unreasonable to work dead and a written request has been made by the main contractor, or his agent. Suitable precautions and testing must also be undertaken before the electrical contractor agrees it is safe to energize the circuit.

Many major injuries have occurred when electrical circuits have been energized before installation was complete

Other aspects of electrical site safety

There are many other ways in which you can ensure site safety with regard to electricity.

- Using reduced voltage equipment reduces the risk of injury and the supply voltage should be limited to the lowest needed to get the job done.
- Battery-operated power tools are the safest.
- Portable tools designed to run from 110V centre-tapped-to-earth supply are readily available.
- Using a residual current device (RCD) with 230V + equipment can reduce injury.
- When working near overhead power lines these should be switched off if at all possible.

Use battery-operated power tools - they are safest to use

Use reduced voltage equipment to reduce risk of injury

Portable tools are readily available

Using Residual Current Device (RCD) can reduce injury

If working near overhead power lines they should be switched off

CHECK YOUR KNOWLEDGE

1. **Imagine you have returned to a job with only a few tasks remaining. You have been wearing overalls and a hard hat, but your gear is in the van. The risks don't seem serious and the work will only take a few minutes. Do you have to use your PPE?**

   ☐ a. No – the work will only take a few minutes.

   ☐ b. No – the risks do not seem serious.

   ☐ c. Yes – the PPE should still be used.

2. **Under the Control of Substances Hazardous to Health Regulations (COSHH) – and European law – who has overall responsibility for controlling exposure to hazardous substances in the workplace?**

   ☐ a. Employers

   ☐ b. Employees

   ☐ c. Government Health and Safety Inspectors

3. **What items do you think should be part of a first-aid kit? Select 7 items and add them to the table shown.**

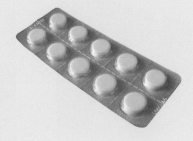

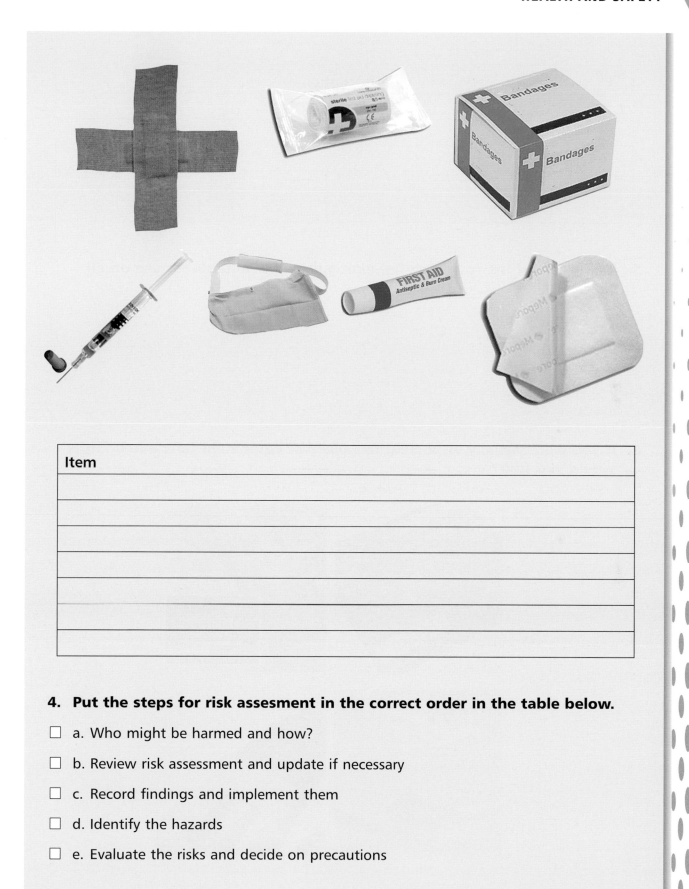

Item

4. Put the steps for risk assesment in the correct order in the table below.

☐ a. Who might be harmed and how?

☐ b. Review risk assessment and update if necessary

☐ c. Record findings and implement them

☐ d. Identify the hazards

☐ e. Evaluate the risks and decide on precautions

Step	Description
1.	
2.	
3.	
4.	
5.	

5. **Which one of the following fire extinguishers is suitable for use on all classes of fire?**

☐ a. Carbon Dioxide

☐ b. Foam

☐ c. Powder

☐ d. Water

6. **If you were visually inspecting this piece of equipment how many faults would you find?**

☐ a. One

☐ b. Two

☐ c. Three

☐ d. Four

7. Match up the fire extinguishers shown with the class of fire they are used on.

Class of Fire	Fire Extinguisher
Class A Wood paper, textiles, etc	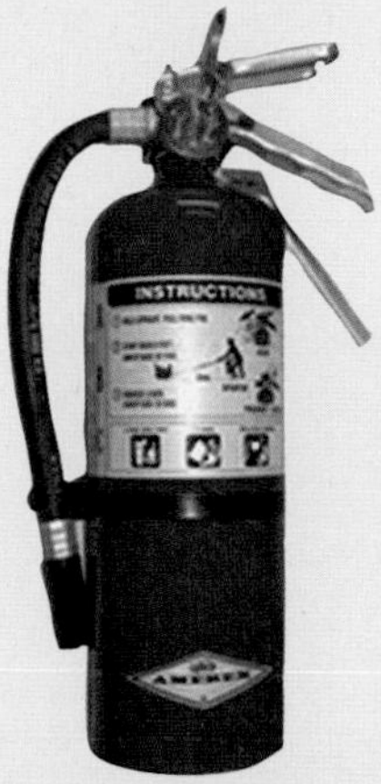
Class B oil, petrol, paint, etc	
Class C gas, acetylene, butane, etc	

| Class D metal, magnesium, aluminium, etc |  |

Chapter 3

HEAT PUMP TYPES

LEARNING OBJECTIVES

By the end of this chapter you will be able to:

- List the different types of heat pump available

- List the components of a heat pump

- List the types of distribution system that can be used with a heat pump

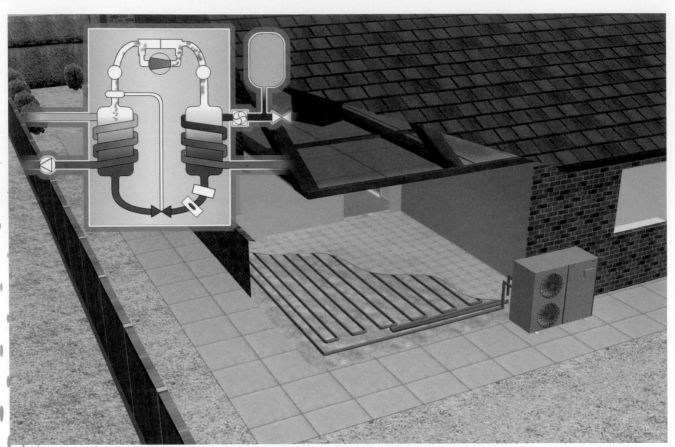

Intersection of a house and illustration of a heat pump

SYSTEM ARRANGEMENTS

There are a number of different types of heat pump systems. These are:

- Outside air to water
- Exhaust air to water
- **Brine** (closed loop) to water
- Water open loop to water
- Direct expansion (DX closed loop) to water

In all of these systems, the collector moves heat from the low temperature heat source via the heat pump to a higher temperature heat sink or distribution system.

Brine The liquid used in ground source heat pumps' collector pipes to absorb heat from the ground. Brine contains a mixture of water and antifreeze.

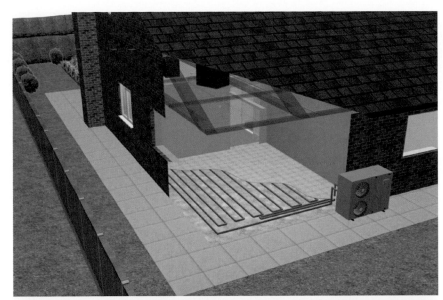

Air source heat pump

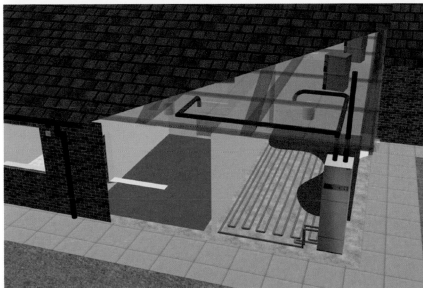

Exhaust air heat pump

Ground Source Heat Pump (GSHP)

Water source heat pump

Direct expansion (DX) heat pump

Heat pump definitions

There is always a debate as to what is a heat pump and what is an air-conditioning unit. Both systems use the refrigeration cycle. Heat can be collected and stored in a solid, a liquid or a gas, so in theory any combination of these phases can be used, i.e.:

- Solid to solid or liquid or air
- Liquid to solid or liquid or air
- Air to solid or liquid or air

In practice, 'to air' systems are typically classed as air-con systems and 'to water' systems as heat pumps, although depending on whether the system is heat only, cooling only or reversible, this heating or cooling preference can also be used to define the system.

Air-conditioning systems

Air-con systems are often 'split' units, meaning the collector (evaporator) is externally located and the distribution system, which often consists of cassettes, fan-coils or, more recently, chilled beams, is located in the various cooled rooms of the building. The refrigerant is then used as the transfer medium and the air-con engineer has an 'F gas' qualification so that he is qualified to fill, commission and empty the circuit with the refrigerant gas.

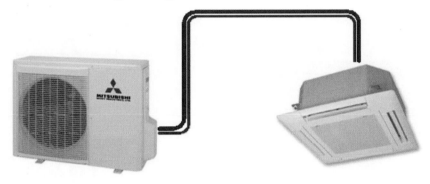

Air-conditioning systems

Types of system arrangement

The majority of the market is for small to medium sized heat pumps up to 100kW for packaged units. These are factory assembled and tested and come complete with all the components and controls.

Type of system arrangement

Ground source heat pump (GSHP)

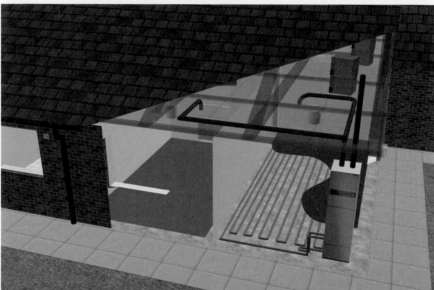

Exhaust air heat pump

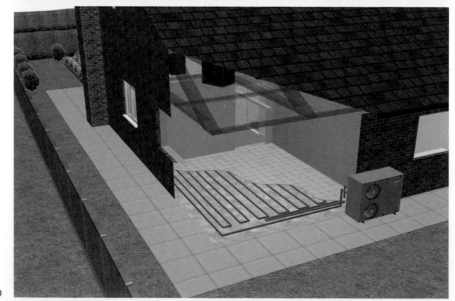

Air source heat pump

Direct expansion (DX) heat pump

Air source heat pump (ASHP)

An air source heat pump or ASHP extracts heat directly from the outside air. This low grade heat is transferred to the heat pump refrigerant. Compression leads to a temperature rise in the refrigerant and this higher grade heat is transferred through a heat exchanger to the heating system water, which can then be used for underfloor heating, low temperature radiators or domestic hot water.

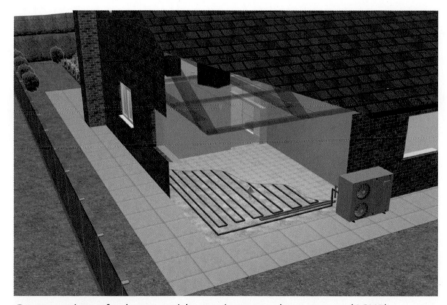

Cross section of a house with an air source heat pump (ASHP)

Exhaust air heat pump

An exhaust air heat pump extracts air from the warm areas of a building, for example the bathroom or kitchen, the heat is removed from the air and transferred to the heat pump's refrigerant circuit before the exhaust air is expelled to the outside. The compression cycle of the heat pump raises the temperature of the refrigerant and transfers it to a water-based system that can be used for space heating or domestic hot water.

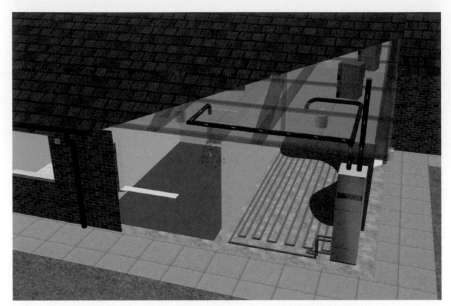

Cross section of a house with an exhaust air heat pump

Ground source heat pump (GSHP)

A ground source heat pump (GSHP) extracts low temperature heat from the ground through pipes which are laid either horizontally in trenches or vertically in boreholes. A water and antifreeze mix known as brine is circulated through a closed loop system of pipes, where it absorbs the low temperature heat. This heats up a refrigerant, which is compressed to raise the temperature. This higher temperature heat is then transferred through the condenser to the heating system water, which can then be used for underfloor heating, low temperature radiators or domestic hot water.

Cross section of a house with a ground source heat pump (GSHP)

Water source heat pump

A water source heat pump extracts heat from water underground which is flowing through a permeable layer of rock. In an open loop system, water is extracted through a borehole which ideally should be upstream of the underground flow of water. This water is passed through the heat pump and re-injected through a second borehole to the downstream flow of the water. The boreholes should be at least 15m apart to avoid the same water being re-circulated.

Cross section of a house with water source heat pump

Direct expansion (DX) heat pump

In a direct expansion (DX) heat pump system, the ground loop is filled with a liquid refrigerant rather than brine. These systems are more thermally efficient than brine ground loops but are more difficult to design. In addition, the metal collection pipes are more prone to damage than plastic and the risk of refrigerant leakage is significantly higher.

Cross section of a house with direct expansion (DX) heat pump

E-LEARNING

Use the e-learning programme to learn more about the different types of system arrangement.

ACTIVITY 5

We know that in winter, a DX heat pump is the most efficient, followed by a GSHP and finally an ASHP. Write a short passage explaining first of all why DX heat pumps contravene environmental legislation in several (but not all) countries, and then why a GSHP is more efficient than an ASHP in winter, and why an ASHP is seen as more efficient than a GSHP in summer.

HEAT PUMP COMPONENTS

The heat pump cycle

The typical heat pump cycle for either an ASHP or GSHP starts with a collector. This is a heat exchanger where the low temperature heat from the water or brine mixture in the ground loops, or the outdoor air, is transferred to either the antifreeze mixture or in some cases, directly to the heat pump's refrigerant. This low temperature heat is then transferred to the refrigerant in the heat pump's evaporator.

The refrigerant gas from the evaporator is then compressed to increase the pressure and temperature of the gas. The condenser is another heat exchanger where the higher temperature heat from this gas is transferred to the distribution system water. As the refrigerant gas loses its heat it condenses back into a liquid. Finally, an expansion valve lowers the pressure and temperature of the refrigerant. This complete cycle of a heat pump circuit is then repeated continuously.

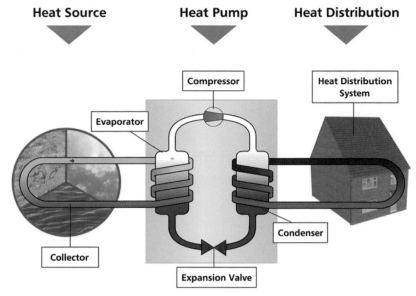

| Heat Source | Heat Pump | Heat Distribution |

Heat pump cycle

Latent heat of evaporation and latent heat of condensation

Heat pumps employ the energy transfer in the latent heat process to deliver significant quantities of heat. Think about a pan of water boiling on a stove. When the water reaches boiling point, the water turns to steam and there is a massive increase in the volume of the fluid. If you stay there and watch the pan boiling, the water will stay at 100°C and continue to produce steam for several minutes as more heat energy is added to the pan of water, until eventually all the water turns to steam.

In physics terminology, this is called the latent heat of evaporation. A lot of energy is required to change the phase of the water from water to steam and this energy input will happen over a period of time. Latent heat is the energy required to turn the water into steam. Likewise, when the steam condenses back to water, it will release this heat. In the kitchen environment, this will occur when the steam touches cooler surfaces such as the windows. We also observe this effect in a sauna, where a steam sauna feels hotter than

Pan of boiling water

a dry sauna, whilst the thermometer will display the same temperature.

This latent heat is used in a heat pump. In the low temperature, low pressure evaporator side of the circuit, the refrigerant will boil at a low temperature (something like 0°C) and absorb heat. The gas will then be compressed to high temperature (something like 40 to 65°C) and high pressure in the compressor. The gas will release its heat energy in the condensor before returning to low temperature and low pressure through the expansion valve.

Heat pump and circuit components

There are a number of key heat pump and circuit components.

The key components of a heat pump (internal heat pump components) are:

- Evaporator
- Low pressure switch
- Compressor
- High pressure switch
- Condenser
- Dryer/Receiver
- Sight glass
- Expansion valve
- Expansion valve phial
- Refrigerant four-way valve

Key circuit components (external parts) are:

- Brine pump
- Emitter circuit valve
- Fan coil
- Buffer tank

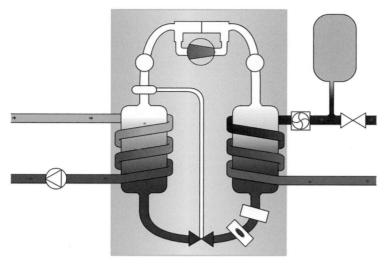

Key components of the heat pump

Evaporator

The evaporator is a heat exchanger transferring heat from the ground or air to the refrigerant. In a GSHP this is usually in the form of a **plate exchanger**, in an ASHP it is a **tube-and-fin exchanger**.

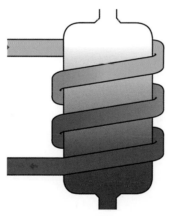

Evaporator

Plate Exchanger A type of heat exchanger that uses metal plates to transfer heat. Plate exchangers are used in ground source heat pumps.

Tube-and-fin Heat Exchanger A type of heat exchanger that uses metal tubes and fins to transfer heat. Plate exchangers are used in the air collector in ASHPs.

Low pressure switch

The low pressure switch sits just in front of the compressor and will detect if the pressure falls too low. In this situation, the switch will operate and the heat pump will go into **fault mode**.

Fault Mode If a heat pump detects a problem it will go into fault mode. The pump may display a message or number specifying the problem.

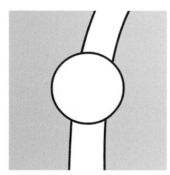

Low pressure switch

Compressor

The compressor takes the low temperature, low pressure refrigerant and raises its pressure. This rise in pressure also raises the temperature of the refrigerant to a useful temperature. The compressor in a heat pump circuit has two distinct functions that are covered in the same device. First of all, it raises the pressure of the refrigerant in its gaseous state and secondly, it also pumps the refrigerant around the circuit and so maintains the flow of the fluid.

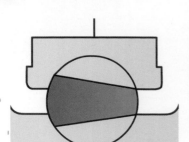

Compressor

The high pressure switch

The high pressure switch sits just after the compressor and will detect if the pressure becomes too high. Like the low pressure switch, it will operate and send the heat pump into fault mode.

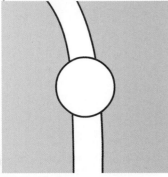

High pressure switch

Condenser

The condenser is also a heat exchanger which transfers the upgraded heat from the refrigerant to the heating system water.

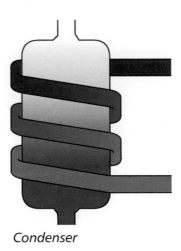

Condenser

Dryer/receiver

Dryer/receiver

The dryer ensures that any remaining moisture or dirt in the refrigerant pipework after the manufacturing process is not circulated around the system.

Sight glass

The sight glass allows the service engineer to see into the refrigerant pipework to check for bubbles or water. Some sight glasses have a lens which changes colour if water is detected in the refrigerant.

Sight glass

Expansion valve

The expansion valve controls the flow of the refrigerant into the evaporator and reduces the pressure. As a result of this drop in pressure the temperature of the refrigerant drops. In most heat pumps the valve adjusts automatically to control the rate of heat transfer and the output temperature.

Expansion valve phial

The expansion valve phial senses the temperature on the output side of the evaporator so the pressure of the refrigerant circuit can be adjusted accordingly.

Refrigerant four-way valve

A four-way valve is added to an air source heat pump to allow the heat transfer process to be reversed. This contributes to the defrost cycle of an ASHP which we will discuss later in this workbook.

Brine pump

The brine pump is located between the **collector loop** and a ground source heat pump and is sized to ensure the minimum flow rate required by the heat pump.

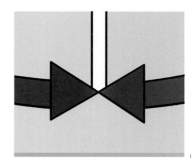
Expansion valve

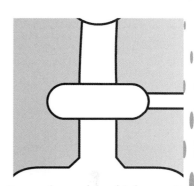
Expansion valve phial

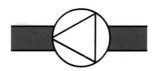
Brine pump

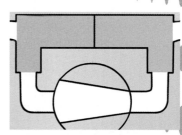
Refrigerant four-way valve

Brine, antifreeze or thermal transfer fluid

The liquid that circulates around the ground collector loop should in strict technical terms be called the thermal transfer fluid. This is because its engineering function is to act as a medium to transfer heat from one place to another. In practice, we tend to use abbreviated terminology on-site. Therefore, a thermal transfer fluid is often called either antifreeze as it is either a pure liquid or an antifreeze and water blend, or brine, representing the salt and water chemical blend of the liquid.

Collector Loop Pipes that are buried in the ground to collect heat energy for use within a ground source heat pump.

Emitter circuit valve

Emitter circuit electromechanical valves are located between the condenser and the distribution system to allow the heated water to flow to either a space heating system, domestic hot water system or both.

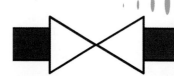
Emitter circuit valve

Y-plan or S-plan

These electromechanical valves are used in most heating distribution circuits to control the movement of the heat. In practice, the industry often uses the term Y-plan to represent a three-way valve that can be adjusted to either provide hot water heating, or space heating only, or with a mid-position, both hot water and space heating. Likewise, two-off two-way valves, often called an S-plan circuit, can be used with either valve open for hot water or space heating, and both open for hot water and space heating. Modern central heating circuits with multiple space heating zones, often controlled from a central **manifold** are often called S-plan plus circuits to represent the multiple space heating zones.

> **Manifold** In relation to heat pumps, the manifold is a component which connects all of the ground collector pipes into a single pipework to feed into the heat pump.

Fan coil

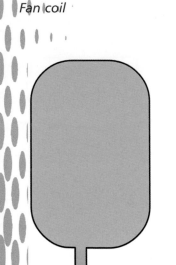

Fan coil

A fan coil is located in an ASHP where the system is an air to air system; it distributes heat by blowing air over the heat exchanger. We will discuss air source heat pumps in more detail later in this workbook. A fan coil can also be used with a GSHP system and this is called a water to air circuit. However, it is not a common application.

Heat and cooling distribution circuits

Heat energy is transferred through heat emitters. These heat emitters can be radiators, underfloor heating or tube-and-fin heat exchangers which use a fan or gravity to move air over the fins. Fan coils, plinth heaters, air curtains and chilled beams are examples of heat exchangers that sometimes use fans and sometimes use gravity convection currents. All heat emitters have advantages and disadvantages and are used in a collection of different scenarios.

Buffer tank

Buffer Tank

The buffer tank is a **thermal store** with high insulation and low heat loss characteristics. It is added to a heat pump circuit to improve the economy and efficiency of the system. However, if there is sufficient water volume in the heating system, a buffer tank is not always necessary.

> **Thermal Store** A form of hot water storage where instead of a large volume of potable tap water being stored a large volume of central heating circuit water is stored and the potable tap water is instantaneously heated via a heat exchanger.

E-LEARNING

Use the e-learning programme to see each of these components.

DISTRIBUTION SYSTEMS

Heating system characteristics

Heat pumps are not designed to be switched on and off like conventional heating systems, they run on a 'tick-over' basis with the pump available continuously to provide heat whenever room temperatures drop below a pre-set level. A system without a buffer tank, normally called an open flow system, is often recommended for underfloor heating circuits with a high water content. Fitting a buffer tank to the heating system is recommended when the heating system has a low water content. This can limit **short cycling** of the heat pump but can increase the cost as compared to an open flow system.

Short Cycling This is when a machine switches on and off regularly in short bursts rather than running more effectively and efficiently over longer time periods.

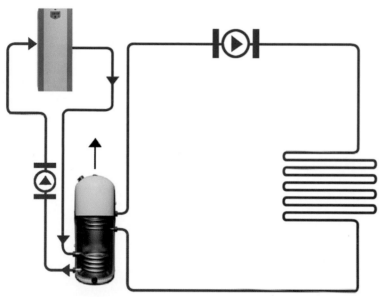

Heating system diagram

Stopping and starting heat pumps

A lot of research and development work is carried out on heat pump circuits. Some heat pumps contain inverters to adjust the flow rate heat pump circuit. Other heat pumps come with soft start features so that they don't draw excessive current loads from the grid during their start-up phase. Individual manufacturers will have their own design strategies for buffering and starting heat pump circuits. Current field trial research indicates that there are several methods for effectively and efficiently designing heat pump circuits and there is no clear indication yet as to a best method. However, there are poor methods such as stopping and starting a heat pump too many times each hour. As more evidence for best practice methods comes onto the market, the industry will provide installers with even more advice for optimizing heat pump circuits. If you follow your manufacturer's or supplier's advice, you should be fitting a high quality heat pump system.

Monovalent systems

A monovalent system is one where the heat pump provides 100 per cent of the heat required by the building. There are various systems available but **environmental** and **economic factors** will play an important part in selecting the correct mode of operation.

Environmental Factors In relation to heat pumps, environmental factors are the potential changes (beneficial or damaging) to the environment when installing a heat pump system.

Economic Factors In relation to heat pumps, economic factors are the financial considerations in regards to the installation and running costs of a heat pump.

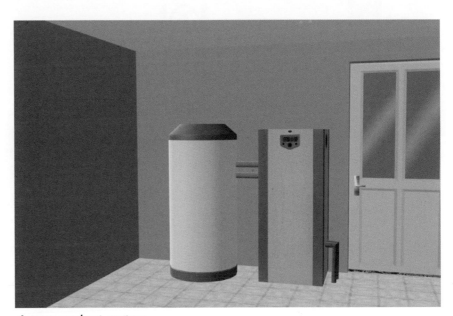

A monovalent system

Space heating only monovalent systems

A monovalent system for space heating only is where a heat pump is providing 100 per cent of the energy required and is linked to a space heating system only such as underfloor heating.

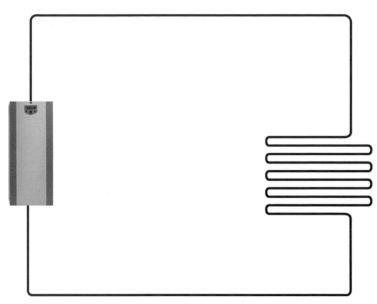

Space heating only monovalent system

Space heating only monovalent systems with a buffer tank

Another type of monovalent system is where a heat pump is providing 100 per cent of the energy required but the system includes a buffer tank in the circuit to prevent short cycling of the heat pump. It is then linked only to a space heating system like underfloor heating.

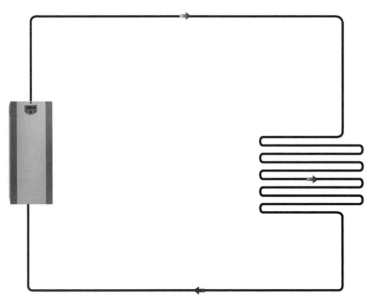

Space heating only monovalent system with a buffer tank

Central heating monovalent systems with a buffer tank on the space heating circuit

A third type of monovalent system is where a heat pump is providing 100 per cent of the energy required and includes a buffer tank in the space heating circuit to prevent short cycling of the heat pump. It is then linked to a space heating system like underfloor heating as well as a domestic hot water system.

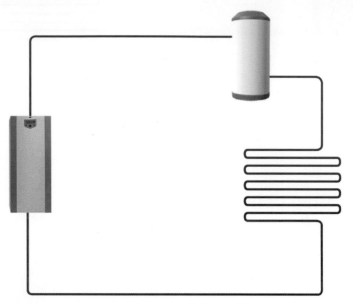

Central heating Monovalent system with a buffer tank

Hot water and space heating with a buffer tank and a solar hot water circuit

Strictly speaking, a hot water and space heating with a buffer tank and a solar hot water circuit is a bivalent circuit, as there are two sources of heat: the heat pump and the solar circuit. However, as the solar energy is free, it is defined as a monovalent system. The heat pump is providing all of the non-solar energy required and in this design, includes a buffer tank in the circuit to prevent short cycling of the heat pump. The heat pump is connected to the upper coil in the hot water cylinder and the solar circuit is connected to the lower coil.

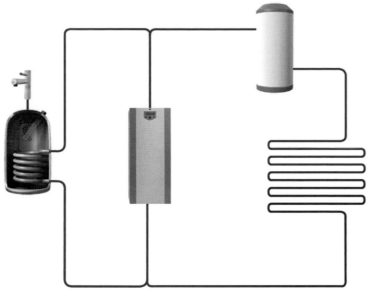

Hot water and space heating with a buffer tank and a solar hot water circuit

Monovalent heat pump connected to a thermal store

The final monovalent system in this section is heating with a thermal store. Thermal stores rely on a large volume of water being heated which can then transfer heat to the domestic hot water via a coil. If carefully sized, these thermal stores can act as a buffer store and so remove the need for a buffer tank.

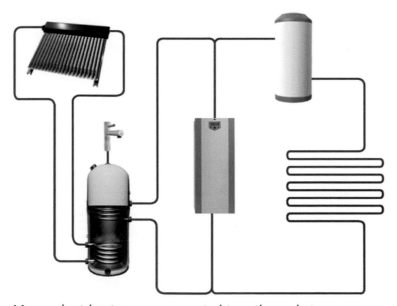

Monovalent heat pump connected to a thermal store

E-LEARNING

Use the e-learning programme to learn more about monovalent systems.

Bivalent systems

Bivalent systems are used when the heat pump cannot provide 100 per cent of the heat required by the building and is split into two types:

- A parallel bivalent system is one where the heat pump contributes to the heating system continuously and is backed up by the supplementary system when it is needed.
- A series bivalent system is one where the heat pump runs until it reaches its maximum capacity, at which point the supplementary system takes over the full load.

Like monovalent systems bivalent systems can include:

- Space heating with buffer tank
- Space and hot water heating with buffer tank
- Space and hot water heating with buffer tank and with a solar system
- Space and hot water heating connected to a thermal store

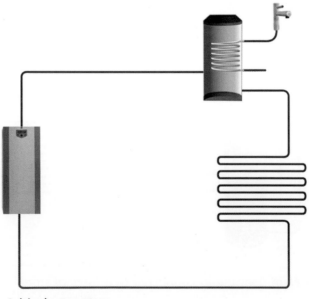

A bivalent system

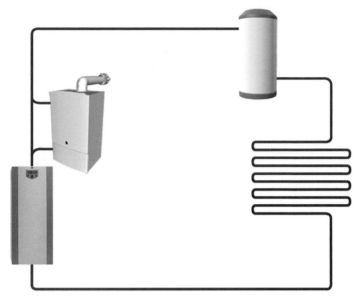

Space heating with a buffer tank

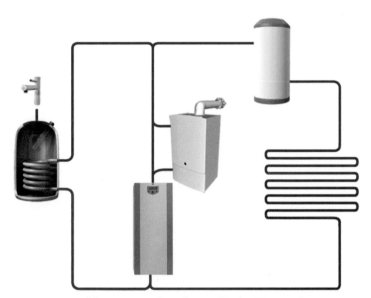

Space and hot water heating with buffer tank

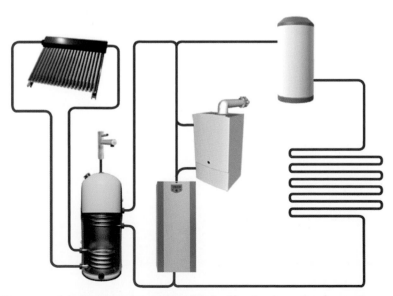

Space and hot water heating with buffer tank and solar system

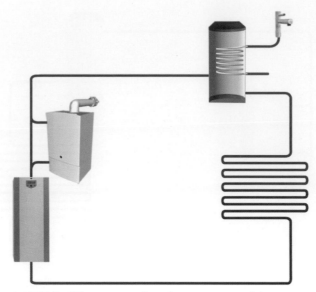

Space and hot water heating connected to a thermal store

ACTIVITY 6

For a thermal store to be effective at providing hot water at 60°C (for Legionella and other bacteria sterilization), the thermal store needs to be kept at a temperature of at least 65°C or even higher so that there is enough temperature differential between the stored primary water and the secondary water delivered to the taps. It is inefficient to produce water at 65°C from a heat pump. Can you suggest any strategies for improving the efficiency of a heat pump with a thermal store?

CHECK YOUR KNOWLEDGE

1. **The image below illustrates the heat pump cycle. Label the different components of the heat pump cycle.**

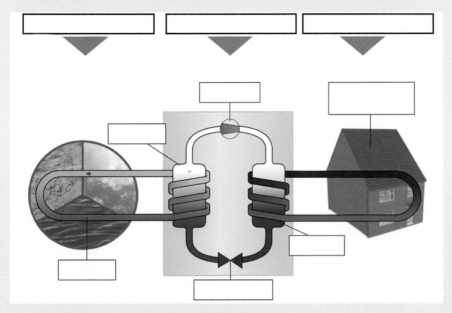

2. **Can you match the different types of heat pump shown with the correct label?**

Air source heat pump (ASHP)	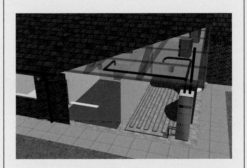
Exhaust air heat pump	

Ground source heat pump (GSHP)	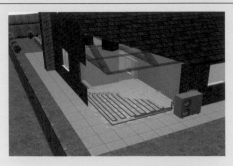
Water source heat pump	
Direct expansion (DX) heat pump	

3. "Heat pumps are designed to run continuously, rather than the shorter cycles that can be used with a conventional boiler." Is this statement True or False?

☐ a. True

☐ b. False

4. What does a buffer tank do? Select 2 answers.

☐ a. Limits the heat pump short cycling

☐ b. Provides additional heat on cold days

☐ c. Increases the volume of water available to the system

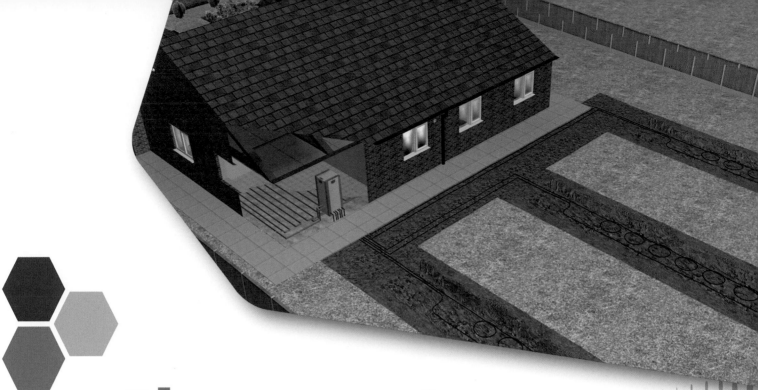

Chapter 4

GROUND SOURCE COLLECTORS

LEARNING OBJECTIVES

By the end of this chapter you will be able to:

- List the different types of ground source collectors

- List the requirements of closed loop ground source collectors

- Explain the basic design principles and component sizing of closed loop ground source collectors

Example of ground source collection

GROUND SOURCE COLLECTORS

Introduction to ground source collectors

Ground source collectors are the means by which heat is collected from the ground and transferred to the heat pump. They consist of lengths of pipe buried in the ground either vertically or horizontally or submerged in water such as a lake. The brine mixture pumped from the heat pump to the collector is typically just below 0°C and because the ground heat is higher than this, the heat from the ground is conducted into the brine mixture, which returns to the heat pump a few degrees higher. This is known as low grade heat.

The flow rate and pipe length are calculated to provide sufficient heat for the building requirements. The pipe length is also limited by the rate at which the heat is extracted from the ground. If the extraction rate is too high, the ground will freeze, reducing the heat available to the collector and reducing the heat pump efficiency.

Types of ground source collector

There are three accepted types of ground source collector:

- Horizontal ground collectors
- Vertical ground collectors
- Surface water collectors

Horizontal ground source collector

Vertical ground source collector

Horizontal ground source collector

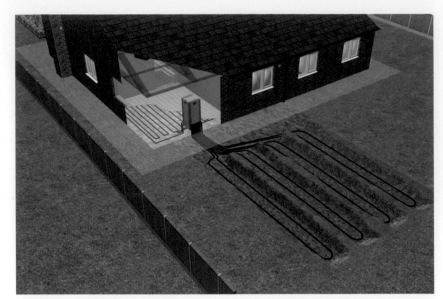

Horizontal ground collector

Horizontal ground source collector

A horizontal ground source collector is typically buried at a minimum of 1m below the ground. This is dependent on there being sufficient land available to lay the required collector area for the building.

Horizontal ground collectors are usually buried at least 1m below ground

Vertical ground source collector

A vertical collector is buried in a borehole and is useful when there is insufficient land for a horizontal collector. Boreholes can be between 15m and 150m deep depending on the heat requirements of the building and the geology of the site. Care should be taken when drilling as it is a specialized process.

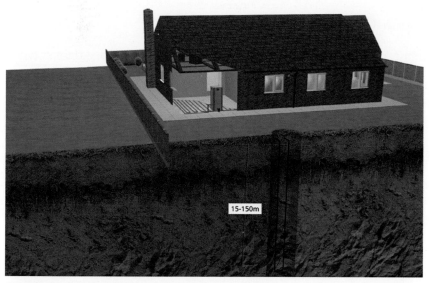

Boreholes can range between 15m and 150m

Surface water collector

A surface water collector is laid at the bottom of a body of water like a lake. The pipes must be weighted to ensure they stay at the bottom of the lake and over a short period of time they will become covered in silt. The size of the lake available is vital to the success of a surface water collector as an insufficient water volume will cause freezing. The water source must also be owned by the same person that owns the building to be heated.

ACTIVITY 7

In theory, all the three collector types listed above could be built completely manually. However, in practice, we use machinery to help us insert these collector types. Thinking about the three different types of collector, what machinery might be used to assist in installing the systems?

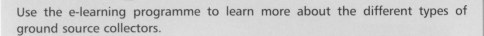

E-LEARNING

Use the e-learning programme to learn more about the different types of ground source collectors.

Surface water collector

ACTIVITY 8

List out one main advantage and disadvantage for each of the above horizontal, vertical and surface water collectors.

Advantage:

Disadvantage:

REQUIREMENTS OF CLOSED LOOP GROUND SOURCE COLLECTORS

Horizontal collectors – linear

Linear horizontal collectors are laid in trenches 1m or more deep. The pipes can be laid in any direction but should typically be spaced approximately 1m apart and 1m from building foundations, services and drainage pipes. For a single pipe, the trench can be 225mm wide and for two pipes it is possible to dig a trench 1200mm wide and lay two pipes side by side with around 1m between them.

A layer of sand or graded soil must be laid below the collector to prevent damage to the pipe from sharp stones with a further layer over the top of the collector once it has been pressure tested. To ensure that no sharp stones or rocks fall into the trench, the next 300mm refill of the trench should be placed and compacted by hand. The remaining back fill can be mechanically placed and compacted.

The surface area above ground source collectors should not be built on or asphalted with a non-porous material as this would prevent the heat from the Sun or rain replenishing the ground heat.

Single pipe - linear horizontal collectors

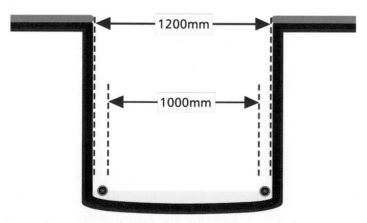

Two pipe - Linear horizontal collectors

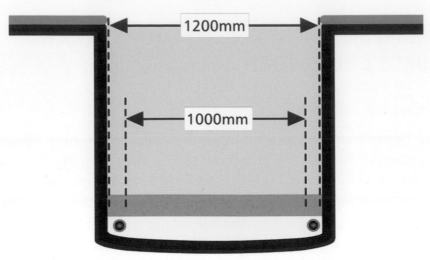

Layer of sand/graded soil must be placed above/below the collector

Horizontal collectors – slinky

Slinky Collectors
Coils of piping filled with a brine mixture buried in trenches to extract heat from the earth.

Because of the increased length of collector pipe, **slinky collectors** extract more heat per unit length of trench than linear collectors. However, as the distance between the trenches is greater, they actually require the same ground area as linear collectors.

Trenches for slinky collectors are typically a minimum of 1200mm wide and 1m deep as well as being 3m to 5m apart. Slinky collectors can also be laid vertically, but the top of the pipe must still be at least 1m deep, meaning the overall trench has to be deeper than for slinky collectors laid horizontally. Slinky collectors can be, if not carefully backfilled, more prone to damage when the trench is backfilled, especially if large voids are left around the pipe.

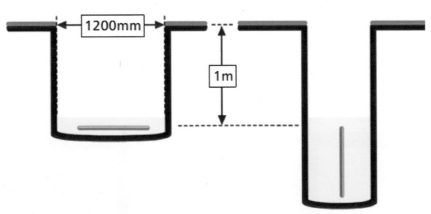

Slinky horizontal ground source collectors

Horizontal collectors – grid

Suppliers often design horizontal grid collectors which use less space than linear or slinky collectors, but a detailed design is vital as these collectors can be prone to freezing the ground due to over-extraction of heat. They are usually supplied prefabricated and are laid either horizontally or on a slope in a trench at least 1m deep with the grid connected in a reverse return layout to balance the flow rates.

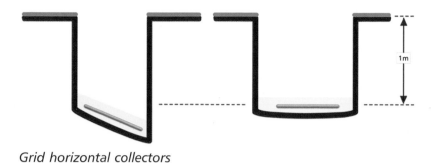

Grid horizontal collectors

Vertical collectors

Where there is insufficient land for a horizontal collector, vertical borehole collectors are used. Boreholes range from 15m to 150m deep depending on the geology of the site and can typically be 6m to 10m apart for domestic buildings.

The pipe for a vertical borehole should typically be 40mm in diameter and have a wall thickness of around 3.7mm. The wall thickness of the pipe accommodates the higher pressure found in a vertical borehole. The pipe is fitted with a u-bend at the bottom of the borehole and a weight is fitted to this u-bend to lower the pipe into the hole.

UK AND INTERNATIONAL STANDARDS

Vertical boreholes are backfilled with a grout called **Bentonite**. This increases the thermal conductivity in the borehole and is generally a requirement of the Environment Agency to prevent cross contamination.

Bentonite An absorbent silicate clay or grout formed from volcanic ash. Bentonite is used to backfill vertical boreholes of ground source heat pumps due to its thermal conductivity and ability to prevent cross contamination.

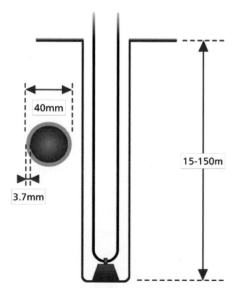

Vertical ground source collectors

Surface water collectors

Surface water collectors in lakes are highly efficient collectors of heat due to the high conductivity of water. The collector in the water resembles a horizontal slinky ground collector but is specially designed by the heat pump supplier. The pipe leading to and from the water is treated in exactly the same way as a horizontal ground collector and is connected to a suitable manifold.

Surface water collector

CIRCUIT DESIGN AND COMPONENT SIZING

Factors affecting heat extraction capacity

There are a number of factors that affect the heat extraction capacity of a heat pump. During a typical season, the ground temperature will vary between a high of 10°C to 16°C and a low of between 3°C and 0°C.

The soil type will also affect the amount of heat that can be extracted; for example, a dry sandy soil will provide approximately 10W/m² whereas a wet clay soil will provide about 40W/m².

A third factor that will affect the heat extraction capacity will be the typical annual operating hours of the heat pump. These can be estimated at eight hours per day from September to the end of April, where the heat pump is providing space heating only, and an additional two hours per day where the heat pump is also providing domestic hot water. In all designs, however, the manufacturer's instructions should always be followed.

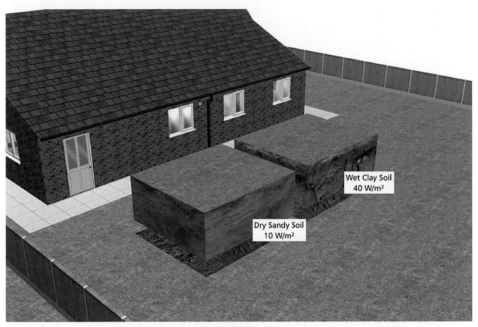

Factors affecting heat extraction capacity

Calculating the land area for a collector

The land (in metres squared) required for a standard collector arrangement is calculated by dividing the output required from the ground collector in watts by the specific heat extraction rate for the soil type in watts per metre squared.

FUNCTIONAL SKILLS

For example, a required output of 7.5kW and a specific heat extraction rate of 25W/m^2 would need 300m^2 of land. This can be affected by the soil conditions so the exact lengths of pipe and land area should be calculated by the supply company. The plot of land can be any shape as long as it covers the correct area.

$$7.5kW = 7500W$$

$$7500W \: / \: 25W/m^2 = 300m^2 \text{ of collector area}$$

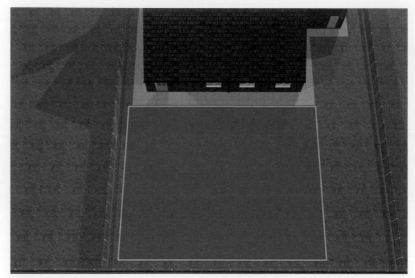

300m² collector area in front of the house

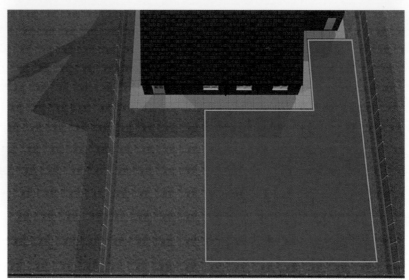

300m² collector area at the side of the house

Calculating linear collector length

Calculating a linear collector length assumes there will be approximately 1m between the pipes. The collector length is calculated by dividing the collector area by the space between the pipes.

FUNCTIONAL SKILLS

For example, if we divide a 300m² land area by a 0.8m space between the pipes, we would need a collector length of 375m. This could be laid as four 47m loops connected in parallel or three 67.5m loops in parallel.

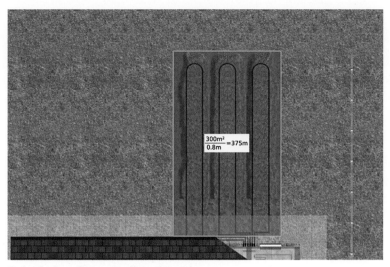

Calculating linear collect length

Calculating slinky collector length

A slinky collector will require the same collector area as a straight pipe collector, however the slinky can be designed with a greater trench spacing. The slinky length is calculated by dividing the collector area by the trench centre line spacing.

FUNCTIONAL SKILLS

For this example, as before, we need a collector area of 300m^2 squared and the collector is designed with 3m between trench centre lines. That's 2m between trenches and a trench width of 1m. If we divide the 300m squared collector area by 3m between trench centre lines, we obtain 100m of slinky. This could be arranged as two parallel trenches of 50m in length or three parallel trenches of 33m in length.

300m^2 / 3m centre line spacing = 100m slinky length

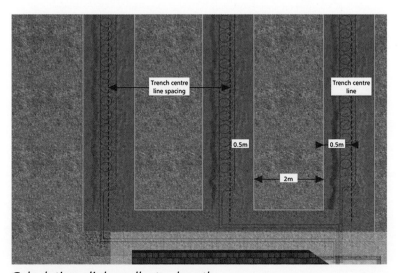

Calculating slinky collector length

ACTIVITY 9

In the above example and in the heat pump e-learning programme, the maths for collector design has been simplified for instructional purposes. In the British MIS 3005 standard, Table 3 sets out a full method for calculating a heat pump collector loop and this standard uses the MIS 3005 supplementary tables.

Using MIS 3005 Table 3 as provided here, fill in Table 3 using the assumptions below:

Table 3 – Details of Ground Heat Exchanger design to be provided to the customer

Parameter	Value		Comments
Estimate of total heating energy consumption over a year for space heating and domestic hot water		kWh [1]	(State calculation method)
HP heating capacity at 0°C ground return temperature and design emitter temperature, H		kW [2]	
FLEQ run hours [1]/[2]		hrs [3]	
Estimated average ground temperature		°C [4]	
Estimated ground thermal conductivity		W/mK [5]	
Maximum power to be extracted per unit length of borehole, horizontal or slinky ground heat exchanger (from the charts and look-up tables), g		W/m [6]	
Assumed heat pump SPF (from heat emitter guide)		[7]	
Maximum power extracted from the ground (i.e. the heat pump evaporator capacity) $G = [2]*1000*(1-1/[7])$		W [8]	
Length of ground heat exchanger calculated using the look-up tables $L_b = [8]/[6]$		m [9]	(i.e. 2 no. 50m slinkies)
Borehole, horizontal loop or slinky spacing, d		m [10]	
Total length of ground heat exchanger active elements, $L_p = [9]*R_{pt}$		m [11]	(NB: does not include header pipes)
Total length of ground heat exchanger active elements installed in the ground, L_p'		m [12]	(NB: state if proprietary software has been used to determine the design length)

Assume the building uses 10 500 kWh for central heating and the heat pump has a capacity of 4.8 kW at 0°C and the design heat emitter temperature is 50°C. The building uses 2400 Full Load Equivalent run hours and you are designing for a horizontal ground heat exchanger. The relevant look-up table and the radiator section of the HE Guide are provided here. Design as if you are in the South West of the UK and the soil conditions are water-saturated wet silt (please note that Rpt for horizontal collectors is 1). Use a flow temperature of 50°C for the radiator section of the HE Guide:

Look-up Table for Horizontal Ground Heat Exchanger at 2400

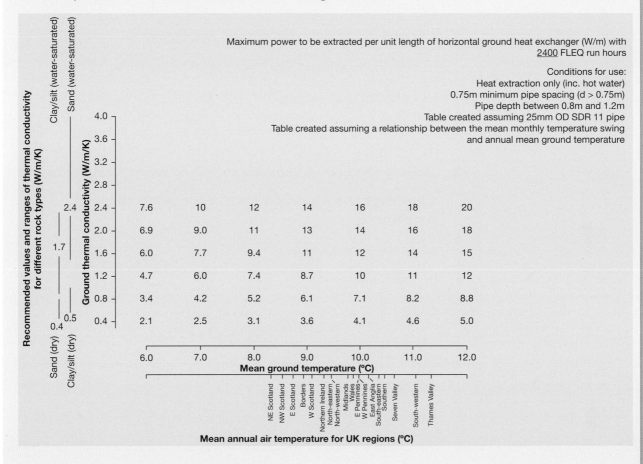

Radiator Section of the Heat Emitter Guide

Use the chart and the calculated Oversize Factor to determine the Temperature Star Rating for that space:

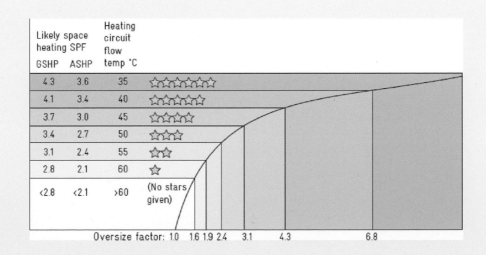

CHECK YOUR KNOWLEDGE

1. **How is the land area for ground source collectors calculated?**

    Complete the equation shown by using these 3 options:

    ☐ a. Output required from ground collector (W)

    ☐ b. Specific heat extraction rate for soil type (W/m^2)

    ☐ c. Collector area required

    $$= \underline{\hspace{5cm}}$$

2. **How is the ground source collector length calculated?**

    Complete the equation shown by using these 3 options:

    ☐ a. Space between pipes

    ☐ b. Land area

    ☐ c. Collector length

    $$= \underline{\hspace{5cm}}$$

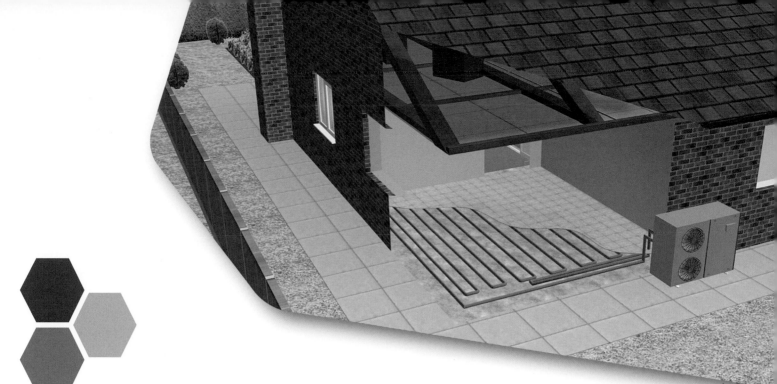

Chapter 5

AIR SOURCE COLLECTORS

LEARNING OBJECTIVES

By the end of this chapter you will be able to:

- List the design considerations that are specific to air source heat collectors

- Explain the design options to provide the defrost cycle for air source heat collectors

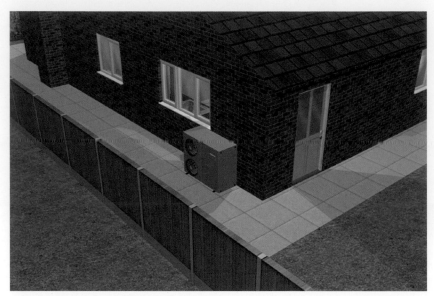

Air source heat pumps

AIR SOURCE HEAT PUMPS

Introduction to air source heat pumps

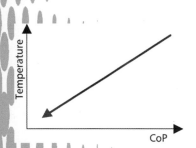

Outside air is a potentially unlimited source of heat but unfortunately the lower weather temperatures in winter reduce the efficiency of an air source heat pump when the heat is most needed. A drop in the outside air temperature will result in a drop in the CoP of the heat pump. A bivalent system may be required if the air source heat pump is not able to meet the total heat requirement of the building.

Types of air source heat pump

An air source heat pump (ASHP) works like a ground source heat pump but extracts heat from the outside air rather than the ground to boil the refrigerant in the evaporator.

Outdoor air systems can be fitted as reversible split systems, which means they can provide cooling as well as heating; the indoor and outdoor air units being linked by refrigerant pipes which run through the wall. However, these reversible systems are not as efficient and cooling does not qualify for government support.

There are also packaged air systems where the outdoor air is ducted to the indoor package. Air source heat pumps have some advantages over ground source heat pumps due to their relatively easy installation. The lack of groundwork for the brine circuit is a particular advantage.

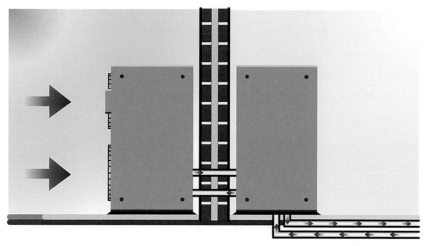

Reversible Split System

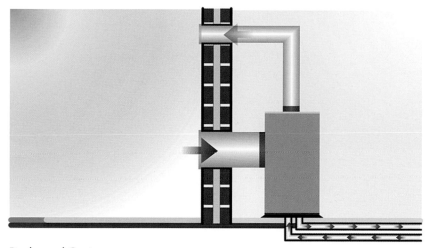

Packaged System

DESIGN CONSIDERATIONS

Siting an air source heat pump

As an air source heat pump uses the outside air as its heat source, anything that affects the temperature of the outside air will affect the efficiency and output of the heat pump, for example the wind speed can affect the speed of the fan in the air unit. An area that is protected from the wind would be preferable in a house that may have a tunnel effect created by a fence at the side of the house. Conversely, the positioning of the ASHP in relation to other structures can create dead areas with little or no wind.

Frost Pocket A low lying area where cooler air gathers and frost occurs more frequently.

The geographical area can also have an effect as the further north you are, generally the cooler it is. North or south facing units will also vary in their efficiency. Finally, the height of the location can affect the heat pump. If it is located lower than the general local ground level, in what is called a **frost pocket**, it will be less efficient.

Siting an air source heat pump

ACTIVITY 10

A house is located on the side of a hill in a rural part of Scotland. The back door faces straight out into the back garden and the windiest section of the hill, straight into the prevailing wind. There is a passageway marked off with a fence down the side of the house. Where might be the best location for the ASHP?

Noise considerations for air source heat pumps

UK AND INTERNATIONAL STANDARDS

The outdoor fan unit is the main cause of noise in an ASHP and is usually more of a problem for the neighbours rather than the occupants of the house with the ASHP installed. Fan noise can be defined as a nuisance under **statutory regulations** so steps should be taken to reduce the noise levels. In a new build this may be part of the planning application, although for wall mounted units there is little that can be done apart from facing the unit away from adjacent houses. With freestanding units there are more options.

> **Statutory Regulations** Licence or approval for engaging in certain situations, e.g. the noise created from an air source heat pump.

The following methods can be used to help reduce noise pollution from an ASHP.

Air source heat pump (ASHP)

- Lawns and plants can act as sound absorbers.

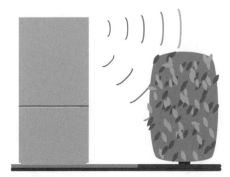

Plants can absorb sound

- Constructed barriers such as walls and fences can reduce noise levels.

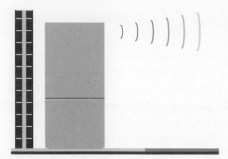

Wall & fences can reduce noise levels

> **Baffle Plate** In relation to air source heat pumps, a baffle plate acts as a barrier which diverts/ dismisses the noises produced by the air source heat pump.

- A **baffle plate** can be used to reduce sound levels.

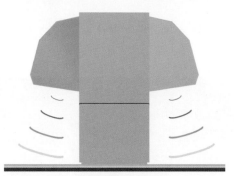

Baffle plate

- Acoustic barriers can be built but these may affect the performance of the ASHP.

Acoustic barriers

- Sitting the ASHP in the valley of a pitched roof can be a possibility.

ASHP placed in a valley of a pitched roof

E-LEARNING

Use the e-learning programme to see more information about reducing noise levels.

ACTIVITY 11

At the start of 2012, to obtain permitted planning development in England and Wales, an ASHP has to be quieter than 42dB at 1m from the neighbour's window. This requirement might be relaxed over time and the requirement might be different in other countries. If you were advising a Government Minister on the types of property most likely to qualify: detached, semi-detached, terrace or flats, what would you tell them?

THE DEFROST CYCLE

Introduction to the defrost cycle

All air contains water, which we refer to as humidity. The amount of water held in the air varies from 15g of water per kilogram of dry air at 20°C, to1.6g of water per kilogram of dry air at −10°C. As air cools and hits the cool part of the evaporator, the water in the air becomes saturated and creates condensation. The point at which condensation forms will vary depending on the outside air temperature, but if this condensation is allowed to freeze, it will cause the heat pump to fail because of low pressure.

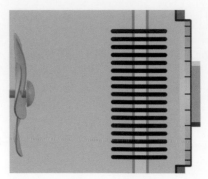

Condensation

The defrost cycle

There are two methods used to overcome the problem of frozen condensation in the heat pump. The first is to include small electric heating elements in the pump. The second and more commonly used design option is to temporarily reverse the heat cycle so that the warm air inside can defrost the evaporator.

The defrost cycle can take between 30 seconds and ten minutes depending on how much ice has built up. In an air to water system it is often necessary to have a buffer tank included in the system. This means there will be sufficient water in the distribution system to allow the defrost cycle to complete.

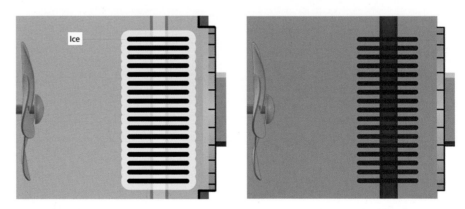

The defrost cycle

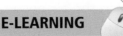

E-LEARNING

See the e-learning programme for more information on the defrost cycle.

ACTIVITY 12

You are asked to specify an ASHP on a property which contains low volume radiators and microbore pipework. You realize that this system has a low water content and so will only have a small 'buffer' to continue heating the house during defrost cycles. In order of increasing effectiveness, briefly explain what the options are for increasing the water content volume of the heating circuit. There is more information on heat emitter circuits in the next chapter so it might help to read this chapter before attempting this activity.

CHECK YOUR KNOWLEDGE

1. **What factors can affect the efficiency of an ASHP? Select which answers you think are correct.**

   ☐ a. Geographical area

   ☐ b. Height of the location

   ☐ c. Positioning of the ASHP in relation to other structures

   ☐ d. Wind speed

2. **Draw the 5 methods which can be used to help reduce noise pollution from an air source heat pump (ASHP).**

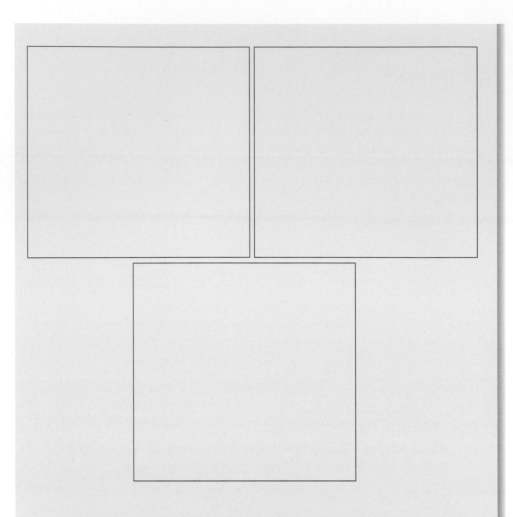

3. **Match up the name of the air source heat pump (ASHP) type with the relevant description and the appropriate image.**

Name	Description	Image
Outdoor air system or reversible split system	This system allows the outdoor air to be ducted into the indoor package.	
Packaged air system	They provide cooling and heating; both units are linked by refrigerant pipes running through the wall.	

Chapter 6

SYSTEM DESIGN

LEARNING OBJECTIVES

By the end of this chapter you will be able to:

- List the types of distribution systems suitable for installation with heat pumps

- Understand the basic principles to calculate the required size of a heat pump

- Understand the implications of installing an oversized or undersized heat pump

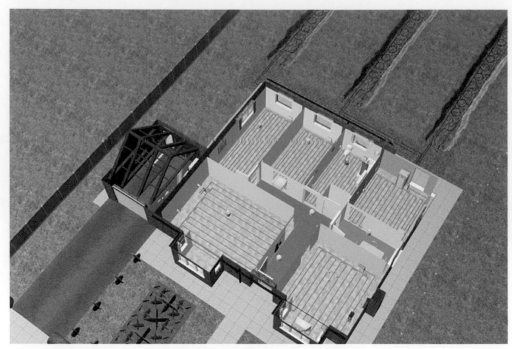

Cross section of a house showing distribution systems

TYPES OF DISTRIBUTION SYSTEM

Introduction to distribution systems

There are three main types of distribution system that can be added to a heat pump circuit:

- Radiators
- Underfloor heating
- Fan-assisted convector heaters.

It is part of the system design to select the most appropriate solution. The decision will be predominantly influenced by the flow or output temperature the distribution system is required to deliver and it should be remembered that the higher this temperature, the lower the efficiency of the system. The mean water temperatures for each system are:

- Radiators – 60°C
- Fan-assisted convectors – 40°C
- Underfloor heating – 35°C

Please note that these are mean water temperatures that typically occur with each system. Different heat emitters can be designed for different output temperatures and will be discussed further in this chapter.

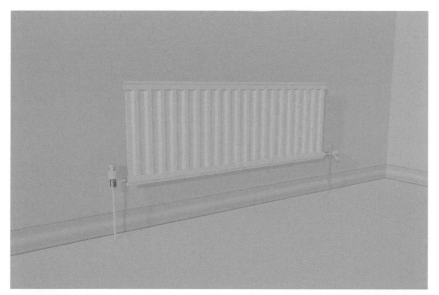

Radiator

Fan assisted convector

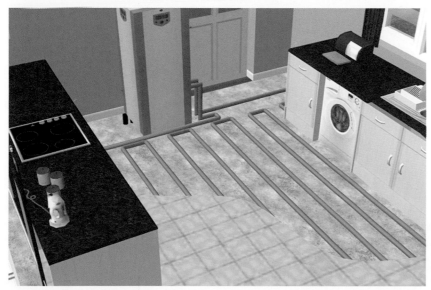

Underfloor heating

Radiators

Radiators can be successfully used as the distribution system with a heat pump, although due to lower flow temperature, the radiators need to be increased in size to achieve the required heat output for the room. This is calculated once an accurate heat loss calculation for the building has been carried out.

Radiator

ACTIVITY 13

Download the Heat Emitter Guide from the MCS website. To achieve a four star temperature rating, how much larger than 'standard' radiator sizes will the radiators need to be? What is the flow temperature at this 4-star setting and what SPF will the system be likely to achieve with a GSHP and an ASHP? Reading the accompanying notes in the HE Guide, what actions are likely to be needed to achieve a 4-star rating with radiators?

Return Temperature
The temperature of the water returning to the source of heat, in this case the heat pump.

FUNCTIONAL SKILLS

Let's take an example where the system has been designed to give a flow temperature of 50°C and a return temperature of 40°C which gives a mean temperature of 45°C. The system is being designed to give a room temperature of 21 degrees Celsius.

To calculate the **Delta T** for this system we subtract the room temperature 21°C from the mean temperature 45°C which gives us a value of 24°C. We would now need to consult the radiator manufacturer's **correction factor** guide to find out the factor for this difference between the mean water temperature and the required room temperature. In this case it is 0.406. Let's assume that our heat loss calculation for this room is 0.8kW. We now need to divide the heat loss figure by the correction factor which gives us a required radiator output of 1.97kW.

A radiator would now be selected that would give us an output of 1.97kW in a conventional system with flow temperature of 75°C which should now give us an output of 0.8kw with the lower temperature flow rate.

$$\text{Delta T} = \text{Mean Water Temperature} - \text{Room Temperature}$$

$$\text{Delta T} = 45°C - 21°C = 24°C$$

$$\text{Correction factor for } 24°C = 0.406$$

$$\text{Heat loss for room} = 0.8kW$$

$$Radiator\ Output = \frac{Heat\ Loss(kW)}{Correction\ Factor}$$

Using standard sizing charts, size of radiator required = 1.97kW

Delta T The difference in temperature between any two points in a circuit, often the difference between the flow and return temperatures or the difference in temperature between the water flowing into and out from a radiator etc.

Correction Factor A set of values provided by manufacturers; the correction factor value is used to divide the heat loss of a room (kW) in order to calculate the radiator output (size of radiator). Specific values are determined by the Delta T or temperature difference value.

ACTIVITY 14

There are many different methods for calculating the output from radiators. In the previous example two methods, oversize factors and correction factors, have been used. All the systems require comparisons between temperature differences. Some are based on flow temperature and others on mean water temperature across the radiator. If the radiator has:

1. A flow temperature of 55°C and a return temperature of 51°C, what is the mean water temperature?
2. A mean water temperature of 40°C and a flow temperature of 43°C, what is the return temperature?
3. A mean water temperature of 60°C and a return temperature of 56°C?

Strictly speaking, in engineering terms, temperature differences should be quoted in Kelvin (K) rather °C. However, in practice, for simplicity, we use °C instead of K. Kelvin is the absolute rather than relative temperature scale. More information on these temperature scales can be found on the Internet. You will see K used in some units such as **U-Values** for heat loss.

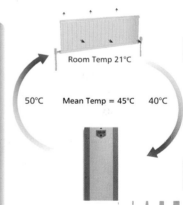

Room Temp 21°C

50°C   Mean Temp = 45°C   40°C

Mean temperature of 40 degrees Celsius

U-Value The amount of heat that passes through a solid surface such as a wall, window or door. The higher the u-value, the greater the heat loss.

Underfloor heating

Underfloor heating is an ideal distribution system for use with a heat pump. It can be the most efficient due to the low temperatures required by the underfloor heating system, which average about 40°C. It is also helped by the large amount of water held in the system, which achieves the minimum flow rate of the heat pump without the need for a buffer tank. The system should be designed to work with a heat pump, which might require less space between the heating pipes than for a conventional installation.

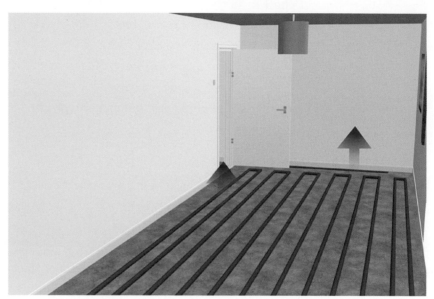

Underfloor heating

Fan-assisted convector heaters

If you are retro-fitting a heat pump into an existing heating system, fan-assisted convector heaters offer another solution. They operate at much lower temperatures than conventional radiators and require less space but depending on location, can be unattractive. A fan-assisted convector system will usually result in a reduction in the water volume of the system, which may require the installation of a buffer tank to maintain the minimum flow rate of the heat pump.

Fan-assisted convector heater

Appearance

The visual appeal or intrusion of heat emitters has always been an important design issue. Heating engineers want to locate the heat emitters in a prominent location in the room where the emitter can quickly respond to the room's needs. Interior designers want to blend and integrate all the features of the room so that the room has a good ambience. This is why expensive designer radiators are now widely available. However, they can be regarded by some as difficult to clean and create dust traps. Likewise, underfloor heating works best with a tile covering as this is a better conductor of heat than wood or carpet flooring. However, a hard tiling floor in a lounge might not provide the desired room effect as it might be seen as too functional and cool for this setting. As ever with design decisions, discussion and compromise optimizes solutions to meet the end user's requirements.

ACTIVITY 15

Explain why in older properties with single glazing, radiators were normally located under the window whilst in modern double glazed and insulated properties, this is not such a commonly used location.

HEAT PUMP SELECTION

Building efficiency

As we have seen, a distribution system with a heat pump is designed to operate at much lower temperatures than a conventional boiler system. Before considering a heat pump system, it is essential that an accurate heat loss calculation is carried out on the building or the system may fall short of expectations and requirements.

FUNCTIONAL SKILLS

Building efficiency can be calculated using a recognized accurate heat loss calculator software package or a full elemental heat loss calculation.

If the efficiency or U-value of the building is found to be poor, steps should be taken to improve the efficiency of the building through the glazing and insulation before a heat pump system is considered and designed.

Accurate heat loss calculations must be carried out

Loft insulation

Double glazing

Sizing a heat pump

Before the correct size heat pump can be selected, the heat load of the building must be calculated. This heat load is always proportional to the outside temperature in that the colder it is outside, the more energy will be needed to keep the building at the required temperature. When calculating the size of the heat pump you need to take into account how low the outside temperature will fall and how often this will happen during the year.

If a heat pump is undersized, it typically means there will be increased consumption of energy as inhabitants heat the building using other means such as electrical heating.

Heat pump in the snow

Chart example 1

This example shows the power required to keep a room at 20°C. If we design a system that assumes a minimum outside temperature of 2°C, we can see that we will need a 5kW heat pump to achieve this. A system designed to this temperature will need to be a bivalent system to allow for periods when the temperature drops below design temperature of 2°C.

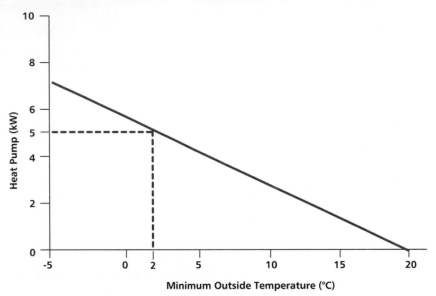

Relationship between heat pump (KW) and minimum temperature outside

Chart example 2

In this example, the system has been designed to a minimum outside air temperature of −3°C which will require a heat pump of 6.8kW. In order to avoid oversizing the heat pump, accurate temperature data will be required to determine how often and how low the external temperature may fall in a given area.

If a heat pump is oversized, it means that installation costs will be increased and it can also cause short cycling.

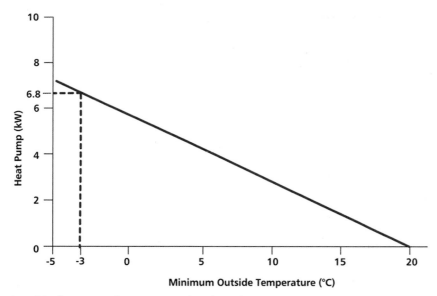

Relationship between heat pump (KW) and minimum temperature outside

Bivalent and mono-energetic

Different definitions are used by different parties in the heating industry. Bivalent and mono-energetic are two terms widely used within the heat pump sector. Bivalency is normally used to describe a system that uses a combustion boiler (gas, wood, coal or oil) in association with a heat pump. A mono-energetic system is normally defined as a heat pump that also uses electrical heating to support the space heating requirements.

ACTIVITY 16

A house requires 0kW heat input to maintain its internal temperature at 20°C outside temperature and 10kW of heat input to maintain its internal temperature at −5°C outside temperature. Use the simple graph below to work out the required kW heat input to maintain the internal temperature at 0°C outside temperature. If the design temperature is 0°C, will the heat pump shown with the power curve below have enough capacity to keep the building warm at 0°C?

Continued on next page

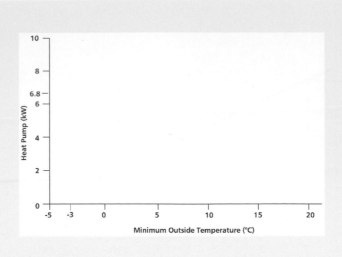

CHECK YOUR KNOWLEDGE

1. **How is the output calculated for a radiator to be used with a heat pump?**

   Complete the equation shown by using these 3 options:

   ☐ a. Correction factor

   ☐ b. Radiator output

   ☐ c. Heat loss (kW)

   = _____

2. **Fan-assisted convector systems may need a buffer tank?**

   ☐ a. True

   ☐ b. False

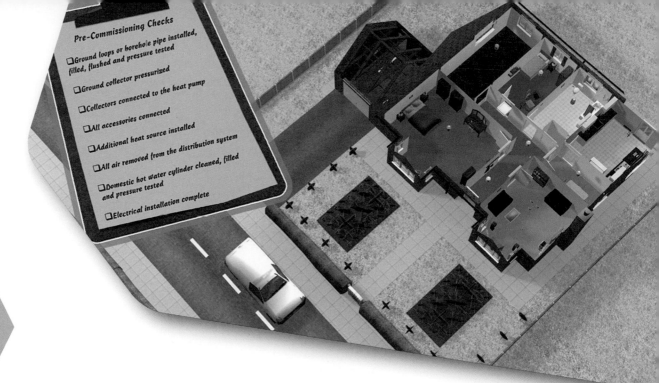

Pre-Commissioning Checks

☐ Ground loops or borehole pipe installed, filled, flushed and pressure tested

☐ Ground collector pressurized

☐ Collectors connected to the heat pump

☐ All accessories connected

☐ Additional heat source installed

☐ All air removed from the distribution system

☐ Domestic hot water cylinder cleaned, filled and pressure tested

☐ Electrical installation complete

Chapter 7

INSTALLATION

LEARNING OBJECTIVES

By the end of this chapter you will be able to:

Explain the preparatory work required prior to a heat pump installation

- List the requirements of a heat pump installation

- Complete a checklist of all installation steps prior to testing, commissioning and handover

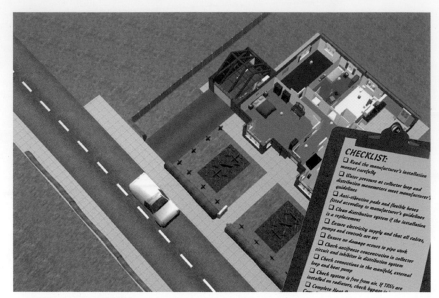

Installation checklist

PREPARATORY WORK

Pre-installation checks

Prior to installing a GSHP or ASHP, a number of pre-installation checks should be carried out:

Pre-installation checks

- Do you have the appropriate permission to proceed with the installation?

Permission checks for installation

- Have you verified the suitability of the **heat distribution system** for connection to the heat pump?

Horizontal ground source collector

Heat Distribution System A system for the transfer of heat around a building. Heat is typically distributed through space heating radiators or under-floor heating.

- Have you verified that the **heat output capacity** of the heat pump unit matches the required proportional contribution to the total heating load of the building? Or in other terms, will this heat pump (and any other additional sources of heat) keep the house warm at the design temperature?

Heat Output Capacity The maximum amount of heat a heat pump can produce.

Heat pump

● Have you verified that the buffer tank size, if fitted, is correct? Or is there enough water volume in the existing heat distribution system?

Buffer tank and heat pump

● Have you checked that you have appropriate access to all the required work areas of the installation?

Satellite view of house

● Have you checked the availability and condition of a suitable electrical input service and informed the utility of the heat pump connection?

Meter reader

● Have you checked that there is adequate provision for the siting of the key internal system components?

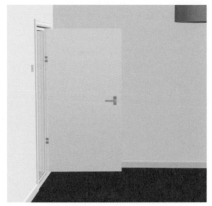

Is there adequate provision for key components?

● Have you checked the suitability of the building structure in relation to the proposed installation?

Cross section of a loft

● Have you carried out the pre-installation checks specific to the positioning of an ASHP, especially noise and clean air flow considerations?

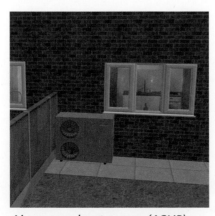

Air source heat pump (ASHP)

ACTIVITY 17

You are located in a rural area and have a design temperature of –2°C specified. The local electric utility company will allow you to fit a 12kW heat pump on the property. At the design temperature of –2°C, the property has a heat loss of 15kW. The 12kW heat pump models you have looked at all have an output of 11.5kW at –2°C. You are short on heating capacity in the building. How could you specify the system to meet the 15kW heat load at –2°C? There is no mains gas supply.

LICENCES AND PERMISSIONS

UK AND INTERNATIONAL STANDARDS

For a heat pump installation, a number of licences and permissions may be required.

Installers should check if your local authorities require any of the following:

- Environment Agency Licence
- Planning Permission
- Building Regulations
- F-Gas Regulations

Licences and permissions required

Environment Agency

A licence may be required from the Environment Agency for some borehole collectors. Although no water is being extracted, the heat in the water is being extracted from the ground which may have an effect on the temperature of the water course and therefore on the local environment.

Environmental Agency for the borehole collectors

Planning permission

Planning permission is not usually required for a heat pump installation unless the works are in a Site of Special Scientific Interest, National Park or Area of Outstanding Natural Beauty. They may be required for an ASHP installation if the noise is likely to cause a disturbance.

You may require planning permission

UK AND INTERNATIONAL STANDARDS

At the start of 2012, in England and Wales, for an ASHP there is a requirement to make sure the heat pump is less than 42dB and 1m from the neighbour's nearest window, for the system to be classed as permitted development, i.e. no requirement for planning permission. This fairly onerous requirement is subject to regular review. Other countries are likely to require similar standards and as ASHPs become more common and noise levels and their consequences become better understood, the local authorities are likely to choose limits that best suit the requirements of the local community.

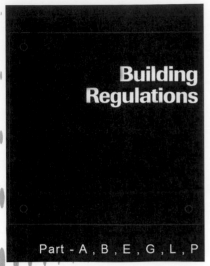

Building regulations Part A, B, E, G, L and P

Fluorine A highly reactive and poisonous gas.

F-Gas certification

Building regulations

Every installation must comply with local building regulations and the installer should be fully aware of these. In England and Wales, this would include part A structure, part B fire, part E transmission of sound, part G hygiene, part L energy efficiency and part P electrical safety. Each country will have its own version of building regulations and they are likely to require a similar set of requirements. For example, building regulations make sure that the building is safe, quiet, efficient, structurally sound and hygienic after the new heating system is fitted.

F-Gas regulations

F-Gas regulations are designed to limit the amount of **fluorine**-containing gases that may be released into the atmosphere, either deliberately or by accident, which may contribute to climate change.

Although F-Gas certification is not required for a sealed heat pump installation, there are some requirements that are specific to heat pump installations. Regular checks should be made for leakage and automatic leak detection systems should be included on large installations. The recovery of the refrigerant, during servicing of the system and at the end of its life should be carried out correctly. Accurate records must be maintained for systems containing 3kg or more of fluorinated gases.

Only personnel with the correct qualifications should be employed during an installation. Other actions also apply, including the labelling of new equipment and ensuring adequate room volume in case of refrigerant leaks.

E-LEARNING

Use the e-learning programme to learn more about licences and permissions.

ACTIVITY 18

A GSHP is fitted in a box (small garden shed) outside the property and is connected via two pipes through into the heating system via a kitchen wall. Please discuss which building regulations from structure, fire, sound transmission, hygiene, energy efficiency and electrical connection might apply. Are there any other licences or permissions that might be required?

Electrical connections

UK AND INTERNATIONAL STANDARDS

All electrical work associated with a heat pump installation must conform to the latest edition of the **IEE** (UK) or National **Wiring Regulations**. In addition a new circuit or one that has been altered will require an Electrical Installation Certificate. The installation of a heat pump can use a large proportion of the available electrical capacity in a building and the system may need to be upgraded as part of the installation.

Heat pumps require an independent circuit back to the local distribution board or consumer unit. The **local District Network Operator**, also called the electric utility company, will need to be informed that a motor, in this case the heat pump compressor, has been permanently connected to their supply.

IEE Wiring Regulations A set of British regulations containing key information regarding the installation of electrical components. IEE Wiring Regulations are also known as BS 7671.

Local Distribution Network Operator (Local DNO) The company responsible for the supply of electricity in the local area.

HEALTH AND SAFETY

All work should be carried out by a suitably qualified electrician.

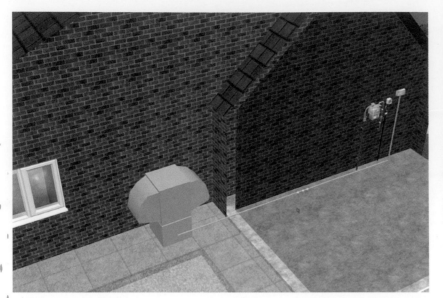

Electrical installation

Requirements for electrical installation

INSTALLATION REQUIREMENTS

Delivery and positioning

When designing a heat pump system, the proposed access routes for installation should be considered. It should also be remembered that these access routes will need to be maintained for the life of the installation to allow repairs and replacements to take place.

HEALTH AND SAFETY

The size and weight of the unit plus any hot water tanks need to be taken into consideration. This is to ensure that they fit through any doors and that the appropriate manual handling guidelines are followed to prevent injury to installers and damage to the heat pump.

Packaged heat pump units for residential installations are usually well soundproofed and can be positioned in kitchens, utility rooms or garages, whereas commercial systems will normally be installed in a plant room. In either case, there should be sufficient space around the system for maintenance and ventilation.

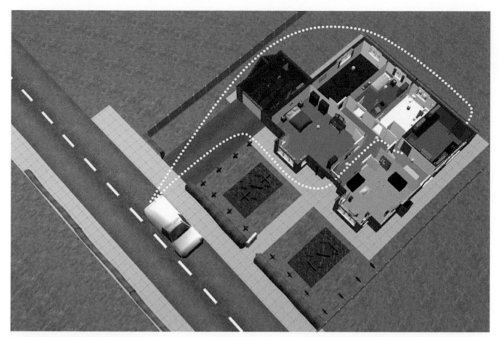

Delivery and positioning

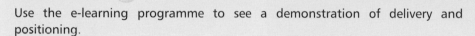

E-LEARNING

Use the e-learning programme to see a demonstration of delivery and positioning.

Manifold positioning

The open ends of all the ground loop collectors in a system are brought together in a manifold configuration which connects all the flows and returns together. There are two options for manifold positioning in a heat pump system. Installation on an internal or external wall is the most common method for a domestic system, whereas an inspection chamber is generally used for commercial installations.

To prevent the ground freezing and to allow for any ground movement that may occur as a result of the installation, the pipes must be insulated a minimum of 1.5m from the entry point of the building. The entry points must also be sleeved against ingress of water and to prevent condensation.

Cross section of a house, showing a manifold configuration

Commercial Inspection Chamber

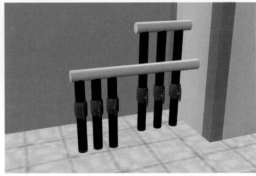

Pipes insulation

ACTIVITY 19

In a previous Activity, we discussed locating a GSHP in a small garden shed. Now think about this property being located in a conservation area and the authorities have denied permission for an external shed. The property has a garage and a reasonably sized kitchen. Explain the advantages and disadvantages of locating the heat pump unit in the kitchen or the garage.

INSTALLATION CHECKLIST

Introduction to the installation checklist

As each installation is unique, it is essential that the manufacturer's installation instructions are followed for installing the collector circuit, the heat pump and the distribution system. However, what follows is a checklist of installation points that would apply to most installations.

In each installation, follow the manufacturer's instructions closely for:

- The collector circuit
- The heat pump
- The distribution system.

House cross section, showing the collector circuit, heat pump and distribution system

Installation checklist

Installation checklist

1. At the start of the installation, you must read the manufacturer's installation manual carefully.

2. Ensure that the water pressure at the collector loop and distribution system **manometers** meet the manufacturer's guidelines.

3. Anti-vibration pads and flexible hoses must be fitted according to the manufacturer's instructions to prevent noise and vibration.

4. If the installation is a replacement the distribution system must be flushed clean. The system should be drained hot, then refilled, then

Manometer An instrument which measures the pressure in collector loops and distribution systems.

Refractometer An instrument for measuring the antifreeze concentration of a brine mixture.

Inhibitor A liquid added to the central heating distribution system water which helps to maintain the heat pump's efficiency.

Thermostatic Radiator Valve (TRV) A self-regulating valve fitted to heating radiators. Room temperature is controlled by the regulation of the flow of heated water to the radiator.

drained again until the water is clean (using a power flusher or system cleaner can assist with this process).

5. Ensure that there is a proper electrical supply, for example 230 volt fused, and that all valves, pumps and other controls are correctly set.

6. Take great care with pipe connections to ensure that no damage is caused to the internal and external pipework.

7. Ensure that the antifreeze concentration in the collector circuit is correct by using a **refractometer**, and that an **inhibitor** has been added to the distribution system. The antifreeze must contain manufacturer specified levels of biocide.

8. Check with a refractometer the antifreeze concentration in the collector circuit and the inhibitor in the distribution system.

9. Check the tightness of all connections in the manifold, external loop and the heat pump before firing.

10. The system must be free from air before firing. If **thermostatic radiator valves** are installed on any radiators ensure that a bypass is installed to maintain the required flow.

11. Finally, complete the Heat Pump Benchmark Commissioning Checklist or that any other manufacturer supplied checklist is completed.

ACTIVITY 20

On commissioning the heating circuit, there are cold spots in some radiators and the pressure gradually drops over time. New oversize radiators have been fitted to the existing pipework circuit. No water, liquid or leaks are evident anywhere on the heat emitter circuit. What might have happened and what could you do to re-commission the circuit?

Chapter 8

TESTING, COMMISSIONING AND HANDOVER

LEARNING OBJECTIVES

By the end of this chapter you will be able to:

- List the requirements to test a heat pump system

- List the requirements to commission a heat pump system

- List the requirements to handover a heat pump system

Front view of a bungalow

TESTING

Pre-testing checks

Prior to testing a heat pump system, a series of checks needs to be carried out.

- Does the heat pump system comply with the system design and specification?
- Does the heat pump system comply with the system and component manufacturer requirements?
- Is the electrical supply circuit arrangement suitable?
- Has the system been flushed of installation debris?
- Have the hydraulic circuits been filled and vented correctly?
- Has the system been protected against freezing?

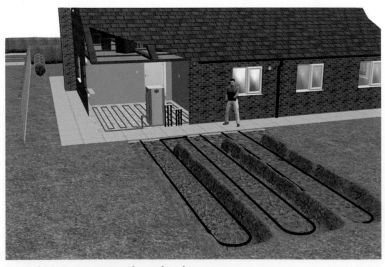

Carrying out pre-testing checks

Pressure testing

A ground source collector must have a certificate or record to show that is has passed a pressure test to accompany any installation. Any defects found later will be expensive and time consuming to repair and would impact on the performance of the system.

After flushing the pipework with a purge pump to remove any debris, pipes are pressure tested with water, either one at a time or with all the pipes connected to the manifold. If a drop in pressure is found, then each pipe needs to be tested individually so that the leak or weak joint can be found.

Pressure testing should be carried out immediately before backfilling and again immediately after. It may also need to be repeated before filling the system if there has been a delay between the backfilling and commissioning of the system.

Pressure testing

ACTIVITY 21

Modern collector circuits are normally tested to EN 806. Here is a graphical method to cover an EN 806 pressure test:

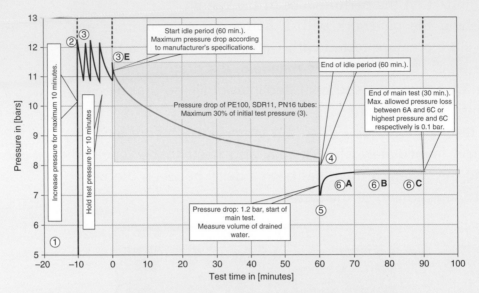

Graphical test procedure (graph from SIA 384/6 (SN 546 384/6)).

Studying the diagram, how long does this test take from start to finish? What equipment will be required to perform the pressure test and what variation in pressure is allowed between the three final test points?

Filling and purging

Purging should be carried out using fresh clean water first, with the antifreeze added at a later stage, according to the purge unit manufacturer's instructions. The purge unit is usually made from a suitable pump connected to a water storage container and needs to have a relatively high flow rate to ensure air is forced out of the collectors and through the purge unit. The flow from the collector to the inlet of the purge unit has a filter which captures any solid material that may be left in the collector to ensure that only clean water is flowing through the collector before it reaches the evaporator.

After purging, the collectors can be filled with the brine mixture. The same purge unit can be used but must be thoroughly cleaned first. Water and antifreeze are mixed to create the brine mixture; ideally this should be created in one batch to ensure even distribution within the collectors. The brine mixture will now have a coloured tint to it and the concentration should be checked using a refractometer after the brine has been circulating in the collectors for around five minutes. After the collector loops have been filled, they should be slowly pressurized to the required point before the flow is shut off.

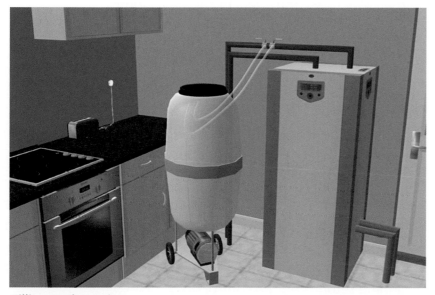

Filling and purging

ACTIVITY 22

The UK standard for fitting heat pumps, MIS 3005, has some specified flow rates for purging the collector circuit. In this purging procedure, the ground collector loop is first purged at a relatively high speed in both directions by itself, followed by a purging of the header and evaporator of the heat pump at a slower speed, with the final purging of the whole collector circuit including array, header and evaporator for at least 15 minutes with a flow velocity of at least 0.6 m/s. Why is the array flushed at a higher speed than the header and evaporator and what advantage is gained by flushing in both directions?

COMMISSIONING

Pre-commissioning checks

Prior to commissioning, a series of checks against the manufacturer's instructions should be carried out.

- Are the ground loops or borehole pipe installed, filled, flushed and pressure tested?
- Is the ground collector pressurized?
- Are the collectors connected to the heat pump?
- Are all accessories connected?
- Has any additional heat source been installed?
- Has all air been removed from the distribution system?
- Has any domestic hot water cylinder been cleaned, filled and pressure tested?
- Is the electrical installation complete?

Setting to work and commissioning of the heat pump can then be completed as with the rest of the installation, by a suitably competent installer in accordance with the manufacturer's instructions. Some manufacturers ask the installer to set the system to work and initially commission the system, and then send in their own engineer for the final commissioning settings on the control unit.

Pre-commissioning checklist

Air source heat pumps

The main checks for an air source heat pump unit are to ensure that airflow rates in the ducted system achieve the required rates and that the noise measurements are appropriate for the installation permissions. The air source heat pump heat distribution system is checked and commissioned in a similar fashion to a ground source heat pump heat distribution system. In addition, some manufacturers ask the competent installer to set the system to work, commission and handover and then supply their own engineer to carry out some fine tuning of the system.

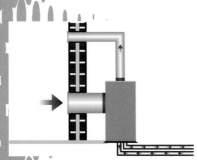

Ducted Air Flow System

Ground source heat pumps

Commissioning of a ground source heat pump also involves balancing the collector loops and checking the overall flow and return rates. It may take a number of days for the air to be fully vented from the system and the pressure may need to be adjusted during this period.

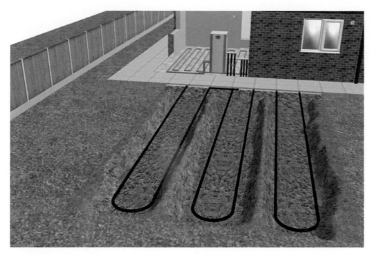

House intersection - process of commissioning a ground source heat pump

HANDOVER

Instructions to client

In order to avoid post-installation problems, the following information should be given to the client on handover of the new heat pump system. The client will need to understand the way the system heats the house; the times of operation and how it differs from a conventional system; whether the system is mono or bivalent and how it will work; the frequency of the stop starts of the system; the expected contribution that the system will make to the household demands; how to boost or override the system and how to set it back to normal operation. Finally, how to optimize the efficiency of the system.

Handover - instructions to the client

Handover documentation

UK STANDARDS

On handover, there will also be a series of documents to give to the client. These include:

- The benchmark certificate
- The manufacturer's commissioning certificate
- Collector test certificates where applicable
- The schedule of equipment and system schematics
- All licences that were granted as part of the installation
- Building, water and electrical certificates
- Instruction manuals for all parts of the system
- Warranty and guarantee details
- Completed grant and incentive paperwork where applicable

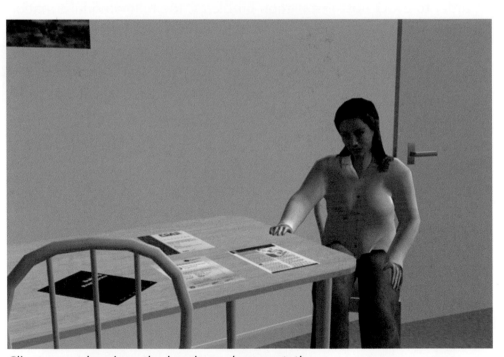

Clients must be given the handover documentation

ACTIVITY 23

In the following heat pump checklist, can you think of anything that might be missing?

HEAT PUMP INSTALLATION CHECKLIST	
Heat Pump Serial Number:	
Customer address:	
Commissioning date:	
	Heat emitter circuit is fully installed, filled, purged, pressurized and holding pressure. Filling loop is disconnected and inhibitor added to this circuit.
	Expansion vessels are sized, fitted and charged in accordance with manufacturer's instruction.
	Pipework is correctly fastened, insulated and wall penetrations appropriately sleeved. Fluid flow indicators as appropriate installed.
	All circuits rise to vent, fall to drain and have vents and drains fitted as appropriate.
	Electrical installation is correct and safe (supply voltage, circuit breakers, earth, etc.) and circuit breakers are installed as recommended by heat pump manufacturer.
	Flows through air/water heat exchangers are in the correct direction.
	All thermostats and sensors are installed correctly and connected to the controller.
	PRV/safety valves operating and overflow pipework flows to drain points.
	Control, TRVs and any other valves are functional and in correct position.
	Heat pump started, operational and sight glass indicating correct operation
	Pump's speed (collector and primary circuit) are set correctly.
	Anti-vibration mounts and flexible hoses installed.
	Temperatures at key points in the system (source flow and return, heating flow and return, refrigerant temperatures at evaporator and condenser) as well as current amp and voltage indicate correct operation.
	Weather compensation, as appropriate, fitted and functioning correctly.
	Ground collector loop pipework layout supplied. Collector loop purged and tested to EN 806.
	Heat pump system commissioned and ready for use.
	All local authority issues such as Environment Agency, planning permission, building control and F-gas regulations followed.
	The Customer/end user has been: • instructed in correct operation of system. • supplied an operation manual • provided with maintenance instructions and schedules • informed of bacteria pasteurization cycle and shown operation of any additional immersion heaters to ensure pasteurization • predicted system performance supplied • predicted running cost including cost for space heating, hot water, additional pumps and immersion heaters supplied.

CHECK YOUR KNOWLEDGE

1. **Brine should be made by mixing water and antifreeze in the collector loops. Is this statement True or False?**

   ☐ a. True

   ☐ b. False

2. **Which of the following should be checked prior to commissioning a heat pump system?**

   ☐ a. Licences have been given to client

   ☐ b. Collectors are installed, filled, flushed and pressure tested

   ☐ c. All air been removed from the distribution system

3. **Once a ground collector has been filled, flushed and pressure tested, the pressure may still need further adjustment. Is this statement True or False?**

   ☐ a. True

   ☐ b. False

Chapter 9

END TEST

END TEST OBJECTIVES

This end test will check your knowledge on the information held within this workbook.

The Test

1. **A system of storage heaters is which type of central heating system?**
   - ☐ a. A dry central heating system
   - ☐ b. A wet central heating system

2. **You are going to be drilling in an extremely dusty environment and it is likely that a lot of noise will be generated. Which of the following items of PPE should you *not* use?**
   - ☐ a. Safety goggles
   - ☐ b. Safety gloves
   - ☐ c. Earmuffs

3. **From where does an exhaust air heat pump extract heat?**
   - ☐ a. From the outside air
   - ☐ b. From the warm areas inside a building

4. **What is a system that uses only the heat pump to provide all the space heating and domestic hot water load called?**
   - ☐ a. Combined
   - ☐ b. Monovalent
   - ☐ c. Bivalent

5. **Which two of the following are bivalent circuits?**
   - ☐ a. Series
   - ☐ b. Combined
   - ☐ c. Parallel
   - ☐ d. Solar

6. **Which of the following are types of ground source heat collector?**
   - ☐ a. Air
   - ☐ b. Horizontal
   - ☐ c. Surface Water
   - ☐ d. Vertical

7. The top of the pipe in a ground source collector should be typically how far below the surface?

☐ a. 0.5m

☐ b. 1m

☐ c. 2m

8. The surface area above ground source collectors can be built on.

☐ a. True

☐ b. False

9. A lower outside air temperature will reduce the CoP of an air source heat pump.

☐ a. True

☐ b. False

10. What are the advantages of an air source heat pump over a ground source heat pump?

☐ a. Bivalent ASHP systems are unnecessary

☐ b. Lack of groundwork needed for ASHPs

☐ c. Relatively easy installation of ASHPs

11. What are the two design options for a defrost cycle on an ASHP?

☐ a. Electrical heating elements in the pump

☐ b. Switch off the heat pump for 30 minutes per day

☐ c. Reverse the heat cycle

12. Which types of distribution system are suitable for use with heat pumps?

☐ a. Fan-Assisted Convector Heaters

☐ b. Radiators

☐ c. Storage Heaters

☐ d. Wet Underfloor Heating

☐ e. Electrical Underfloor Heating

13. What is the mean average water temperature for many types of underfloor heating circuit?

☐ a. 10°C

☐ b. 35°C

☐ c. 60°C

14. Using the chart, what size heat pump (kW) will be required for a minimum outside air temperature of 0°C to keep the room temperature at 20°C?

kW

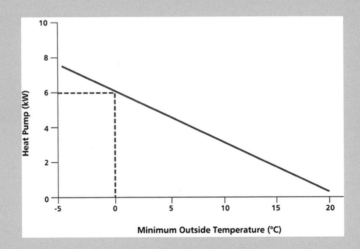

15. Which of the following agencies/permissions may need to be consulted/sought prior to a heat pump installation? Enter a Yes or No in the second column of the table.

Agencies/Permissions	Yes or No
Building Regulations	
Environment Agency	
F-Gas Regulations	
Highways Agency	
Planning Permission	
Gas Utility	

16. Access to the heat pump unit and collector is only necessary during the installation of the heat pump system.

☐ a. True

☐ b. False

17. Which type of manifold position is generally used for domestic installations and which for commercial installations? Enter the correct type in the table

☐ a. Internal/External Wall

☐ b. Inspection Chamber

Domestic Installations	Commercial Installations

18. What is used to measure the concentration of the antifreeze in the collector circuit?

☐ a. Manometer

☐ b. Sight Glass

☐ c. Refractometer

19. When should pressure testing of the collector loops in a ground source heat pump be carried out?

☐ a. Before backfilling

☐ b. During backfilling

☐ c. After backfilling

20. On handover, which of the following are verbal instructions for the client and which are written documentation for the client? Enter the following items in the correct column in the table

☐ a. Benchmark certificate

☐ b. Collector test certificates

☐ c. Completed grant paperwork

☐ d. Electrical certificates

- [ ] e. Expected contribution that the system will make to the household demands
- [ ] f. Frequency of the stop/starts of the system
- [ ] g. How to boost or override the system and how to set it back to normal operation
- [ ] h. How to optimize the efficiency of the system
- [ ] i. How the system heats the house
- [ ] j. Instruction manuals
- [ ] k. Licences
- [ ] l. Manufacturer's commissioning certificate
- [ ] m. Times of operation and how it differs from a conventional system
- [ ] n. Whether the system is mono or bivalent and how it will work
- [ ] o. Warranties and guarantees

Verbal Instructions	Documentation

Answers

1. A – Dry central heating system

2. B – It is dangerous to wear safety gloves if you are using machinery

3. B – An exhaust air heat pump extracts air from the warm parts of a building like the bathroom and kitchen

4. B – A system that uses only the heat pump to provide all the space heating and domestic hot water load is called a monovalent system

5. A, C – Series and parallel are both types of bivalent circuit

6. B, D, C – Horizontal, vertical and surface water are all types of ground source heat collectors

7. B – The top of the pipe in a ground source collector should typically be 1m below the surface

8. B – False, the surface area above ground source collectors should not be built on or asphalted as this would prevent the heat from the Sun or rain replenishing the ground heat. The only exception to this would be if the surface was porous

9. A – True, a drop in the outside air temperature will reduce the CoP of the pump

10. B and C – The lack of groundwork for the brine circuit and the relatively easy installation are the main advantages of ASHPs over GSHPs

11. A, C – Electrical heating elements in the pump or, more commonly, temporarily reversing the heat cycle are both design options for a defrost cycle on an ASHP

12. A, B, D – Fan-assisted convector heaters, radiators and wet underfloor heating are all suitable for use with heat pumps

13. B – The mean average water temperature for many types of underfloor heating circuit is 35°C

14. 6 – For a minimum outside air temperature of 0°C, a 6kW heat pump will be required to keep the room at 20°C

15. Yes = A, B, C, E No=D, F – Building Regulations, F-Gas Regulations and Planning Permission may need to be consulted/sought prior to a heat pump installation. For some types of heat pump, the Environment Agency may also need to be consulted

16. B – False, access to the heat pump unit and collector must be maintained for the life of the installation to allow repairs, replacement and maintenance to take place as well as allowing ventilation of the system

17. A = Domestic Installation, B = Commercial Installation – Domestic installations generally use a manifold attached to an internal

or external wall while commercial installations which tend to be larger have the manifold in an inspection chamber

18. C – A refractometer is used to measure the concentration of the antifreeze in the collector circuit

19. A and C – Pressure testing should be carried out immediately before backfilling and again immediately after. It may also need to be repeated before filling the system if there has been a delay between the backfilling and commissioning of the system

20. Verbal Instructions = E, F, G, H, I, M, N – Documentation = A, B, C, D, J, K, L, O

Activity Answers

Activity 1

For the combustion boiler, if it is supplied with pure biogas, bio-oil or wood heat, then the boiler could be said to be 100 per cent renewable. Very few biofuels are 100 per cent renewable as some fossil-fuel energy is normally used in the biofuel's manufacture and distribution.

For the heat pump, if the heat pump is run on 100 per cent renewable electricity, e.g. wind powered or PV powered energy is used to run the pump instead of fossil fuel-derived electricity, the heat pump can be said to be 100 per cent renewable energy. However, it is not an easy or cheap process to ensure completely renewable electricity.

Activity 2

The ASHP will be the best option for the summer only swimming pool heating system. This is because the air is warmer than the ground during the summer months and so it provides a higher temperature heat source.

The swimming pool heating system will be more efficient than the hot water heating system as the swimming pool is typically heated to about 29 to 30°C and the hot water system is heated to 60°C. Therefore, the lower temperature of the swimming pool heat sink makes this system more efficient.

Activity 3

The electricity costs 0.15 pence/kWh (= 545.00/3650 kWh)

The customer's predicted use is 9000 kWh for space heating and 3000 kWh for hot water heating (75:25 split on 12 000)

The customer's predicted use is 2432.43 kWh of electricity to provide the space heating (= 9000/3.7)

The customer's predicted use is 1071.43 kWh of electricity to provide the hot water heating (= 3000/2.8)

The customer's predicted use is 3503.86 kWh of electricity to power the GSHP system (=2432.43 + 1071.43)

The customer's predicted spend is (= 525.58 pence/annum to provide the central heating.

Activity 4

Using the HE Guide, by insulating and draught-proofing the property, it was possible to increase the star rating by 2 stars. Similarly, by increasing the radiator size, it was possible to increase the star rating by a factor of 2. In this case, the radiators are already 'oversize' by a factor of 1.6. By increasing this oversize factor to 2.4, the flow temperature could be reduced to 50°C and the SPF would be 2.7 with a 3 star rating. If the insulation and draught-proofing changes were added to the 2.4 oversize factor radiators, a 5 star rating might be realized with a flow temp of 40°C and an SPF of 3.4. To achieve the 6 star rating, the property would typically require both the insulation and underfloor heating. Please also note that this is just general advice. In practice, nothing can be demonstrated without a room-by-room heat loss calculation on the property.

Activity 5

DX heat pumps directly circulate the refrigerant through the ground. Many countries prevent this practice because they don't want refrigerants leaking into the ground and its associated water table. Other countries are OK about handling the risks associated with this policy. GSHP is more efficient in winter because the ground is warmer than the air in winter and the air is warmer that the ground in summer. Because we want space heating in winter, GSHPs are more efficient than ASHPs for space heating, whilst for the constant HW heating load, ASHPs should prove as efficient as GSHPs over the whole year. This can only be confirmed with field trials.

Activity 6

Various strategies could be used. One might be to have two separate cylinders, a large cooler one acting as the low temperature buffer store and hot water preheater with the water leaving this low temperature store and being fed into a high temperature store which is then connected to the points of use. Or the store designer could design the cylinder to have different layers functioning at different temperatures. This layered store is called stratified storage.

Activity 7

The horizontal collector trench might be built with a small JCB or other digger with a scoop on the front. A vertical trench might be built with a trenching machine dragged behind a tractor.

The vertical hole would be built with a drilling machine and these come in many different shapes and sizes for the many different geological conditions that can be encountered. A surface water collector would often be weighted and sunk to the bottom of a lake and so a small boat can assist in this process.

Activity 8

Horizontal. Advantage – relatively easy and so lower cost to install. Disadvantage – because it is near the surface, it is more prone to interference from surface temperature and being dug up.

Vertical. Advantage – obtains a lot of heat from a relatively small area and because of its depth below 15m, it can also obtain heat from the Earth's core as well as solar heat from the Earth's surface. Disadvantage – drilling is more expensive than trenching and so vertical collectors cost more.

Surface water. Advantage – lowest cost of all the above. Disadvantage – probably the least efficient as the lake will probably be colder than the ground in winter and also most properties don't typically have a large enough body of water adjacent to the property.

Activity 9

Parameter	Value			Comments
Estimate of total heating energy consumption over a year for space heating and domestic hot water	**10,500**	kWh	[1]	(State calculation method)
HP heating capacity at 0°C ground return temperature and design emitter temperature, H	**4.8**	kW	[2]	
FLEQ run hours [1]/[2]	**2400**	hrs	[3]	
Estimated average ground temperature	**11**	°C	[4]	
Estimated ground thermal conductivity	**1.7**	W/mK	[5]	**Typical figure for Wet Silt**
Maximum power to be extracted per unit length of borehole, horizontal or slinky ground heat exchanger (from the charts and look-up tables), g	**14**	W/m	[6]	
Assumed heat pump SPF (from heat emitter guide)	**3.4**		[7]	
Maximum power extracted from the ground (i.e. the heat pump evaporator capacity) $G = [2]*1000*(1-1/[7])$	**3388**	W	[8]	**= 4.8 * 1000 * (1 - 1/3.4)**
Length of ground heat exchanger calculated using the look-up tables $L_b = [8]/[6]$	**242**	m	[9]	(i.e. 2 no. 50m slinkies)
Borehole, horizontal loop or slinky spacing, d	**0.75**	m	[10]	**taken from look-up table**
Total length of ground heat exchanger active elements, $L_p = [9]*R_{pt}$	**242**	m	[11]	(NB: does not include header pipes)
Total length of ground heat exchanger active elements installed in the ground, L_p'	**242**	m	[12]	(NB: state if proprietary software has been used to determine the design length)

Details of Ground Heat Exchanger design to be provided to the customer

MIS 3005 and its supplementary supporting documents can be found on the MCS website.

Activity 10

If the ASHP is located near the back door, it will be in the direct line of the prevailing wind and so have long defrost cycles. However, if the ASHP is located in the passageway, the ASHP might suffer from air recirculation caused by the adjacent fence. Therefore, if the ASHP is located facing into the back garden but on the side of the house close to the fence, the fence will act as a wind break and so let the heat pump function more effectively.

Activity 11

The only way to accurately assess whether a property passes or fails the 42 dB at 1m from the neighbour's window or similar test is to measure it. These days, a modern smartphone app can be used to check this noise level. However, 42 dB is fairly quiet so you would probably advise the Government Minister that unless the ASHP model is particularly quiet, only detached houses are normally likely to pass this noise level test.

Activity 12

Please note that doing the maths is the only way to demonstrate the actual volume changes that the changes to the system will have. However, in most situations, changing to higher volume radiators and 15mm rather than microbore pipework would add more water content. Wet underfloor heating normally contains more water than radiators and a well sized buffer store will normally add the most water content to the system.

The internal volume of the heat emitter circuit is calculated by working out the internal area of the pipe or cylinder and multiplying it by the pipe length or cylinder height respectively. Radiator suppliers should specify the internal volume of their radiators.

Activity 13

The radiators will need an 'oversize factor' of greater than 3.1 to achieve a 4 star rating. At this rating, the water flow temperature can be 45°C and a GSHP will be likely to achieve 3.7 SPF and an ASHP 3.0 SPF. Regards actions, the room will probably need to be well insulated with good double glazing and draught-proofing and the radiator will probably need to be a double convector design so that it doesn't become overly intrusive of room space and so unsightly appearance.

Activity 14

1. The temperature difference is 4°C and 4/2 = 2.55 − 2 = 53°C which is the mean water temperature.

2. The temperature difference between the flow and mean water temperature is 3°C (43 − 40 = 3). The mean water temperature minus the temperature difference is 37°C (40 − 3 = 37) which is the return temperature.

3. The temperature difference between the mean water temperature and the return temperature is 4°C (60 − 56°C). The flow temperature is 64°C which is calculated by adding the temperature difference to the mean water temperature (4 + 60 = 64).

Activity 15

In older single glazed properties, the window area was the coldest section of the room. The radiator was normally placed under the window to balance out the cold convection currents in the room, so looking to create an even temperature throughout the room. In modern well insulated houses, there is less of a cold spot near the window and so this requirement, whilst still beneficial, is less important.

Activity 16

The house needs 8 kW of heat to maintain its internal temperature of 20°C at 0°C outside temperature. Therefore, this heat pump which produces just under 7kW of heat energy will be too small to maintain the internal temperature of 20°C at 0°C outside.

Activity 17

A monoenergetic system won't work as the electrical supply to the house is obviously already at maximum load to supply the electricity for the heat pump, so any additional load will probably cause tripping at the consumer unit. Therefore a bivalent system is required. Because there is no mains gas supply, you will probably need to look at a bivalent oil, LPG or biomass (woodheat) boiler to back up the heat pump. This bivalent system could be parallel (running the heat pump and combustion boiler at the same time) or series wired so that the heat pump switches off below a certain temperature and the other heat source then covers 100 per cent of the heat load. Some of you might have specified a mCHP system and this could also be an exotic solution although mCHP and some other innovative heat sources are not yet widely used. As stated before, a mains gas boiler won't work as there is no mains gas at this property.

Activity 18

Technically speaking, all building regulations always apply to every system. In practice, we

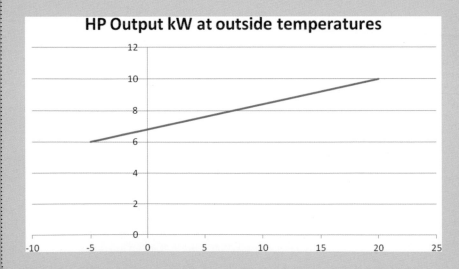

HP Output kW at outside temperatures

need to discuss which regulations need to be reviewed in this scenario. Structurally the building is unlikely to be affected although it might be advisable to fit the HP on anti-vibration mounts. The pipe connections through to the kitchen should be fire-proofed. They could be sleeved in 'rockwool' and sealed with fire-proof mastic either end. Sound with an external GSHP should not be a problem and the HP is specified with anti-vibration mounts. If an indirect cylinder with a heat exchanger is fitted for hot water provision, this change should not affect the tap water and so a hygienic water supply circuit should be maintained. The system should be designed with energy efficiency in mind and with as low a flow temperature as can be feasibly achieved at realistic cost and house intervention. The electrical connection will always have to be made with local requirements taken into consideration and followed. Planning permission on a domestic GSHP would probably not be required although fitting a garden shed to a listed building or in an area of outstanding natural beauty would probably require various permissions. If the GSHP used an open-loop borehole whereby it pumped water up from and re-injected it back into an aquifer borehole, it would probably require permission from the local environment agency. However, a closed loop horizontal or vertical borehole would probably be fine. F-gas regulations normally only apply if the heat pump contains more than 3 kg of refrigerant. If in doubt on any of the above, check rather than assume.

Activity 19

In either location, the HP unit will take up space and it will often be about the size of a kitchen cupboard. It will be far less obtrusive in the garage and less prone to noise and vibration interference. However, it will be, depending on the location of the garage, further away from the property and so potentially less responsive and efficient. As ever, it will be a buyer beware decision and your responsibility is to communicate the options and let the customer decide.

Activity 20

The system has probably not been cleaned out properly and flushed before the new radiators were fitted and dirt has probably moved around the circuit creating blockages, air pockets and cold spots. The system needs to be reflushed, cleaned out and recommissioned and this will hopefully address all the issues. It is important to make sure that all systems, whether brand new or recommissioned, are flushed to remove all debris including fluxes and pastes used during the installation process.

Activity 21

The test takes 100 minutes from start to finish as the test runs from −10 to 90 minutes. The minimum equipment required will be a pump to pressurize the circuit, a manometer (pressure gauge) to measure the pressure and two valves to seal the circuit during the test procedure. The pressure changes between the final three test points must be less than 0.1 bar.

Activity 22

A brine solution tends to entrap small air bubbles and so purging at high speed for some time tends to release these air bubbles. The ground array can be flushed at a higher speed than the header and evaporator as these higher speeds might damage the internals of the header and evaporator. Likewise, flushing in both directions encourages the release of the entrapped air bubbles.

Activity 23

This Heat Pump Commissioning Checklist is fairly comprehensive. However, three other points (and you might have thought of others) could probably be included as follows:

- has an appropriately sized buffer vessel been installed and its use explained to the customer?
- have manufacturer's instructions been followed?
- have local and national requirements been followed? For example, in the UK to be MCS compliant, MIS 3005 must be followed and the documentation table in this standard sets out a series of national requirements including information from the Heat Emitter Guide and on defrost cycles.

Always follow regulations and manufacturer's instructions.

You should also want to supply your own company details on commissioning certificates so as to obtain repeat business and be professional.

Check your knowledge answers

Chapter 1

1. Complete the equation shown by using these 3 options:

$$SPF = \frac{\text{Sum of delivered heat}}{\text{Sum of electricity consumption}}$$

2. Circle the correct letter from the options shown:

A	B	C	D	E	F	G	H	I	J	K	(L)	M
N	O	P	Q	R	S	T	U	V	W	X	Y	Z

Chapter 2

1. Yes – PPE should still be worn.
2. a – You must take responsibility for your safety, however the law requires employers to assess risks, set up control measures and ensure good work practice.
3. Plasters, eye wash, adhesive dressings, sterile eye pads, sterile bandages, sling, sterile wound dressings - you should probably add small items of equipment such as safety pins, disposable gloves, tweezers and scissors. However, please note that tablets or medicines should not be part of a first-aid kit, as if used improperly they could cause further harm.
4. d, a, e, c, b
5. c – Powder
6. c – There are three faults, the fuse does not match the rating plate, the cable has been repaired, and there are burn marks on the casing.

7.

Class of Fire	Fire Extinguisher
Class A – Wood paper, textiles, etc	
Class B – oil, petrol, paint, etc	

Class C – gas, acetylene, butane, etc	
Class D, metal, magnesium, aluminium, etc	

Chapter 3

1.

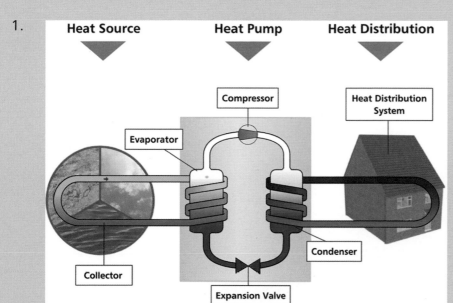

2.

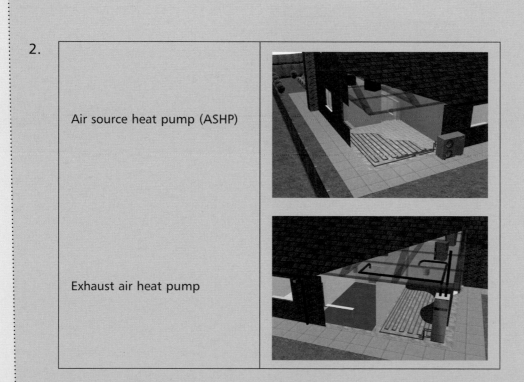

Air source heat pump (ASHP)

Exhaust air heat pump

Ground source heat pump (GSHP)	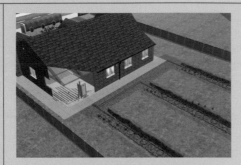
Water source heat pump	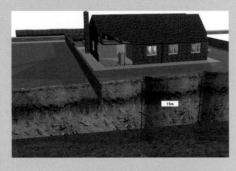
Direct expansion (DX) heat pump	

3. a. True
4. a. Limits the heat pump short cycling
    c. Increases the volume of water available to the system

Chapter 4

1. Collector area required = $\dfrac{\text{Output required from groundcollector (W)}}{\text{Specific heat extraction rate for soil type (W/m2)}}$

2. Collector length = $\dfrac{\text{Land area}}{\text{Space between pipes}}$

Chapter 5

1. a. Geographical area
    b. Height of the location
    c. Positioning of the ASHP in relation to other structures
    d. Wind speed

2.

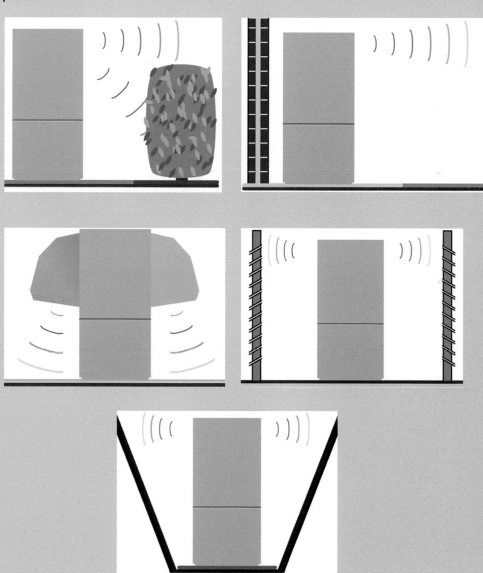

3.

Name	Description	Image
Outdoor air system or reversible split system	They provide cooling and heating; both units are linked by refrigerant pipes running through the wall.	
Packaged air system	This system allows the outdoor air to be ducted into the indoor package.	

Chapter 6

1. Radiator output $= \dfrac{\text{Heatloss (kW)}}{\text{Correction factor}}$
2. a. True

Chapter 8

1. b. False
2. b. Collectors are installed, filled, flushed and pressure tested
   c. All air been removed from the distribution system
3. a. True

Glossary

Air Source Heat Pump (ASHP) A system which extracts heat from (or expels heat to) the outside air to upgrade the heat in a heat pump in order to heat (or cool) a building. Most ASHPs are heat only.

Baffle Plate In relation to air source heat pumps, a baffle plate acts as a barrier which diverts/dismisses the noises produced by the air source heat pump.

Bentonite An absorbent silicate clay or grout formed from volcanic ash. Bentonite is used to backfill vertical boreholes of ground source heat pumps due to its thermal conductivity and ability to prevent cross contamination.

Boreholes A narrow shaft drilled vertically into the ground. These can be used to accommodate vertical collector pipes with a ground source heat pump. Boreholes for ground source heat pumps can range from 5m to 150m deep.

Brine The liquid used in ground source heat pumps' collector pipes to absorb heat from the ground. Brine contains a mixture of water and antifreeze.

Coefficient of Performance (CoP) A measure of the heat pump's efficiency. The CoP is calculated by dividing the heat output of the pump in kilowatts by the electrical input in kilowatts. CoPs can be expressed as a number, ratio or percentage. (e.g. '4', '4:1' or '400 per cent').

Collector Loop Pipes that are buried in the ground to collect heat energy for use within a ground source heat pump.

Correction Factor A set of values provided by manufacturers; the correction factor value is used to divide the heat loss of a room (kW) in order to calculate the radiator output (size of radiator). Specific values are determined by the Delta T or temperature difference value.

Delta T The difference in temperature between any two points in a circuit, often the difference between the flow and return temperatures or the difference in temperature between the water flowing into and out from a radiator etc.

Domestic Hot Water (DHW) Water which is heated and supplied for washing and bathing via taps or showers. DHW is always potable water and should be handled carefully to manage both Legionella and scalding risks.

Economic Factors In relation to heat pumps, economic factors are the financial considerations in regards to the installation and running costs of a heat pump.

Environmental Factors In relation to heat pumps, environmental factors are the potential changes (beneficial or damaging) to the environment when installing a heat pump system.

Fault Mode If a heat pump detects a problem it will go into fault mode. The pump may display a message or number specifying the problem.

Flow Temperature The temperature of the water leaving a source of heat, in this case, the heat pump.

Fluorine A highly reactive and poisonous gas.

Frost Pocket A low lying area where cooler air gathers and frost occurs more frequently.

Ground Source Heat Pump (GSHP) A system that uses heat from the ground to heat (or cool) a building. It uses the earth as a heat source in the winter or a heat sink (in cooling mode) in the summer.

Heat Distribution System A system for the transfer of heat around a building. Heat is typically distributed through space heating radiators or under-floor heating.

Heat Exchanger A heat exchanger is a device designed to efficiently transfer heat from one medium to another.

Heat Output Capacity The maximum amount of heat a heat pump can produce.

Heat Source In relation to heat pumps, the heat source is heat which can be extracted for use within a heat pump. Heat sources for heat pumps include air, ground and water.

IEE Wiring Regulations A set of British regulations containing key information regarding the installation of electrical components. IEE Wiring Regulations is also known as BS 7671.

Inhibitor A liquid added to the central heating distribution system water which helps to maintain the heat pump's efficiency.

Kilowatt (kW) The kilowatt is equal to one thousand watts. This unit is used to express the output power of engines, the power consumption of tools and machines, the heating and cooling power used and any other forms of power.

Local Distribution Network Operator (Local DNO) The company responsible for the supply of electricity in the local area.

Manifold In relation to heat pumps, the manifold is a component which connects all of the ground collector pipes into a single pipework to feed into the heat pump.

Manometer An instrument which measures the pressure in collector loops and distribution systems.

Plate Exchanger A type of heat exchanger that uses metal plates to transfer heat. Plate exchangers are used in ground source heat pumps.

Refractometer An instrument for measuring the antifreeze concentration of a brine mixture.

Return Temperature The temperature of the water returning to the source of heat, in this case, the heat pump.

Short Cycling This is when a machine switches on and off regularly in short bursts rather than running more effectively and efficiently over longer time periods.

Slinky Collector Coils of piping filled with a brine mixture buried in trenches to extract heat from the earth.

Statutory Regulations Licence or approval for engaging in certain situations, e.g. the noise created from an air source heat pump.

Thermal Store A form of hot water storage where instead of a large volume of potable tap water being stored, a large volume of central heating circuit water is stored and the potable tap water is instantaneously heated via a heat exchanger.

Thermostatic Radiator Valve (TRV) A self-regulating valve fitted to heating radiators. Room temperature is controlled by regulation of the flow of heated water to the radiator.

Tube-and-fin Heat Exchanger A type of heat exchanger that uses metal tubes and fins to transfer heat. Plate exchangers are used in the air collector in ASHPs.

U-Value The amount of heat that passes through a solid surface such as a wall, window or door. The higher the u-value, the greater the amount of heat loss.

Index